T0189954

# Advances in Intelligent Systems and Computing

Volume 375

*About this Series*

The series “Advances in Intelligent Systems and Computing” contains publications on theory, applications, and design methods of Intelligent Systems and Intelligent Computing. Virtually all disciplines such as engineering, natural sciences, computer and information science, ICT, economics, business, e-commerce, environment, healthcare, life science are covered. The list of topics spans all the areas of modern intelligent systems and computing.

The publications within “Advances in Intelligent Systems and Computing” are primarily textbooks and proceedings of important conferences, symposia and congresses. They cover significant recent developments in the field, both of a foundational and applicable character. An important characteristic feature of the series is the short publication time and world-wide distribution. This permits a rapid and broad dissemination of research results.

*Advisory Board*



More information about this series at http://www.springer.com/series/11156

Ross Overbeek · Miguel P. Rocha
Florentino Fdez-Riverola
Juan F. De Paz
Editors

# 9th International Conference on Practical Applications of Computational Biology and Bioinformatics

*Editors*
Ross Overbeek
Fellowship for the Interpretation of Genomes
Burr Ridge, IL
USA

Miguel P. Rocha
Department of Informatics, Centre of Biological Engineering
University of Minho
Braga
Portugal

Florentino Fdez-Riverola
Department of Informatics, ESEI: Escuela Superior de Ingeniería Informática
University of Vigo
Ourense
Spain

Juan F. De Paz
Departamento de Informática y Automática, Facultad de Ciencias
University of Salamanca
Salamanca
Spain

ISSN 2194-5357 ISSN 2194-5365 (electronic)
Advances in Intelligent Systems and Computing
ISBN 978-3-319-19775-3 ISBN 978-3-319-19776-0 (eBook)
DOI 10.1007/978-3-319-19776-0

Library of Congress Control Number: 2015940430

Springer Cham Heidelberg New York Dordrecht London


Printed on acid-free paper

Springer International Publishing AG Switzerland is part of Springer Science+Business Media (www.springer.com)

# Preface

Biological and biomedical researches are increasingly driven by experimental techniques that challenge our ability to analyze, process, and extract meaningful knowledge from the underlying data. The impressive capabilities of next generation sequencing technologies, together with novel and ever evolving distinct types of omics data technologies, have put an increasingly complex set of challenges for the growing fields of Bioinformatics and Computational Biology. To address the multiple related tasks, for instance in biological modeling, there is the need to, more than ever, create multidisciplinary networks of collaborators, spanning computer scientists, mathematicians, biologists, doctors, and many others.

The International Conference on Practical Applications of Computational Biology & Bioinformatics (PACBB) is an annual international meeting dedicated to emerging and challenging applied research in Bioinformatics and Computational Biology. Building on the success of previous events, the 8th edition of PACBB Conference will be held during 3–5 June 2015 in the University of Salamanca, Spain. On this occasion, special issues will be published by the Journal of Integrative Bioinformatics, the Journal of Computer Methods and Programs in Biomedicine, and the Current Bioinformatics journal covering extended versions of selected articles.

This volume gathers the accepted contributions for the 8th edition of the PACBB Conference after being reviewed by different reviewers, from an international committee composed of 72 members from 15 countries. The PACBB'15 technical program includes 15 papers from 26 submissions spanning many different subfields in Bioinformatics and Computational Biology.

Therefore, this event will strongly promote the interaction of researchers from diverse fields and distinct international research groups. The scientific content will be challenging and will promote the improvement of the valuable work that is being carried out by the participants. In addition, it will promote the education of young scientists, in a postgraduate level, in an interdisciplinary field.

We would like to thank all the contributing authors and sponsors (Telefónica Digital, Indra, Ingeniería de Software Avanzado S.A, IBM, JCyL, IEEE Systems Man and Cybernetics Society Spain, AEPIA Asociación Española para la

Inteligencia Artificial, APPIA Associação Portuguesa Para a Inteligência Artificial, CNRS Centre national de la recherche scientifique), as well as the members of the Program Committee and the Organizing Committee for their hard and highly valuable work and support. Their effort has helped to contribute to the success of the PACBB'15 event. PACBB'15 would not exist without your assistance. This symposium is organized by the Argonne National Laboratory (USA), the University of Salamanca (Spain), the Next Generation Computer System Group (http://sing.ei.uvigo.es/) of the University of Vigo (Spain) and the University of Minho (Portugal).

Ross Overbeek
Miguel P. Rocha
PACBB'15 Programme Co-chairs

Florentino Fdez-Riverola
Juan F. De Paz
PACBB'15 Organizing Co-chairs

# Organization

## General Co-chairs

Miguel Rocha—University of Minho (Portugal)
Ross Overbeek (Co-Chairman)—Argonne National Laboratory (USA)
Florentino Fdez-Riverola—University of Vigo (Spain)
Juan F. De Paz (Co-Chairman)—University of Salamanca (Spain)

## Program Committee

Alejandro F. Villaverde—IIM-CSIC (Spain)
Alfredo Vellido Alcacena—UPC (Spain)
Alicia Troncoso—University Pablo de Olavide (Spain)
Amin Shoukry—Egypt-Japan University of Science and Technology (Egypt)
Amparo Alonso—University of A Coruña (Spain)
Ana Cristina Braga—University of Minho (Portugal)
Ana Margarida Sousa—University of MInho (Portugal)
Anália Lourenço—University of Vigo (Spain)
Armando Pinho—Universty of Aveiro (Portugal)
Boris Brimkov—Rice University (US)
Carlos A.C. Bastos—University of Aveiro (Portugal)
Carole Bernon—IRIT / UPS (France)
Carolyn Talcott—Stanford University (US)
Consuelo Gonzalo—Technical University of Madrid (Spain)
Daniel Glez-Peña—University of Vigo (Spain)
Daniela Correia—University of Minho (Portugal)
David Hoksza—Charles University in Prague (Czech Republic)
David Rodríguez Penas—IIM-CSIC (Spain)
Ernestina Menasalvas—Universidad Politécnica de Madrid (Spain)

Eva Balsa-Canto—IIM-CSIC (Spain)
Eva Lorenzo Iglesias—University of Vigo (Spain)
Fernanda Correia Barbosa—University of Aveiro (Portugal)
Fernando Diaz-Gómez—University of Valladolid (Spain)
Filipe Liu—University of Minho (Portugal)
Florencio Pazos—CNB/CSIC (Spain)
Florian Leitner—CNIO (Spain)
Francisco Couto—University of Lisboa (Portugal)
Gael Pérez Rodríguez—University of Vigo (Spain)
Giovani Librelotto—Federal University of Santa Maria (Brasil)
Gustavo Isaza—University of Caldas (Colombia)
Gustavo Santos-García—University of Salamanca (Spain)
Heri Ramampiaro—Norwegian University of Science and Technology (Norway)
Hugo López-Fernández—University of Vigo (Spain)
Isabel C. Rocha—University of Minho (Portugal)
Javier De Las Rivas—CSIC (Spain)
Javier Tamames—CNB CSIC (Spain)
João Ferreira—University of Lisboa (Portugal)
João Manuel Rodrigues—University of Aveiro (Portugal)
Joel P. Arrais—DEI/CISUC University of Coimbra (Portugal)
Jorge Vieira—IBMC, Porto (Portugal)
José Antonio Castellanos Garzón—University of Valladolid (Spain)
Jose Ignacio Requeno—University of Zaragoza (Spain)
José Luis Oliveira—Universty of Aveiro (Portugal)
José Manuel Colom—University of Zaragoza (Spain)
Juan Antonio García Ranea—University of Malaga (Spain)
Juha Plosila—University of Turku (Finland)
Julio R. Banga—IIM-CSIC (Spain)
Loris Nanni—University of Bologna (Italy)
Lourdes Borrajo Diz—University of Vigo (Spain)
Luis de Pedro—Autonomous University of Madrid (Spain)
Luis F. Castillo—University of Caldas (Colombia)
Luis M. Rocha—Indiana University (USA)
Mª Araceli Sanchís de Miguel—University of Carlos III (Spain)
Manuel Álvarez Díaz—University of A Coruña (Spain)
Marcelo Maraschin—Federal University of Santa Catarina, Florianopolis (Brazil)
Maria Olivia Pereira—IBB—CEB Centre of Biological Engineering (Portugal)
Mark Thompson—Leiden University Medical Center (The Netherlands)
Martin Krallinger—CNIO (Spain)
Martín Pérez-Pérez—University of Vigo (Spain)
Masoud Daneshtalab—University of Turku (Finland)
Mehmet Tan—TOBB University of Economics and Technology (Turkey)
Miguel Reboiro—University of Vigo (Spain)
Mohammad Abdullah Al-Mamun—Northumbria University (UK)
Mohd Firdaus Raih—National University of Malaysia (Malaysia)

Mohd Saberi Mohamad—Universiti Teknologi Malaysia (Malaysia)
Monica Borda—University of Cluj-Napoca (Romania)
Naresh Singhal—Department of Civil and Environmental Engineering, The University of Auckland (New Zealand)
Narmer Galeano—Cenicafé (Colombia)
Natthakan Iam-On—School of Information Technology Mae Fah Luang University (Thailand)
Nuno F. Azevedo—University of Porto (Portugal)
Nuno Fonseca—CRACS/INESC, Porto (Portugal)
Óscar Dias—CEB/ IBB, Universidade do Minho (Portugal)
Pablo Chamoso—University of Salamanca (Spain)
Patricia González—University of A Coruña, Computer Architecture Group (GAC) (Spain)
Paula Jorge—IBB—CEB Centre of Biological Engineering (Portugal)
Pedro Sernadela—University of Aveiro (Portugal)
Pierpaolo Vittorini—University of L'Aquila (Italy)
Ramón Doallo—University of A Coruña (Spain)
René Alquezar Mancho—UPC (Spain)
Rita Ascenso—Polytecnic Institute of Leiria (Portugal)
Roberto Costumero—Universidad Politécnica de Madrid (Spain)
Rosalía Laza—University of Vigo (Spain)
Rubén López-Cortés—Universidade Nova de Lisboa (Portugal)
Rubén Romero González—University of Vigo (Spain)
Rui Brito—University of Coimbra (Portugal)
Rui Camacho—Universty of Porto (Portugal)
Sara C. Madeira—IST/INESC ID, Lisbon (Portugal)
Sara P. Garcia—University of Aveiro (Portugal)
Sara Rodríguez—University of Salamanca (Spain)
Senay Kafkas—EMBL Outstation Hinxton The European Bioinformatics Institute (UK)
Sérgio Deusdado—Polytecnic Institute of Bragança (Portugal)
Sergio Matos—DETI/IEETA (Portugal)
Sherin El Gokhy—Egypt-Japan University of Science and Technology (Egypt)
Silas Villas-Boas—Univerity of Auckland (New Zealand)
Suhaila Zainudin—National University of Malaysia (Malaysia)
Thierry Lecroq—Univeristy of Rouen (France)
Tiago Resende—University of Minho (Portugal)
Valentin Brimkov—SUNY Buffalo State College (US)
Vanessa Maria Gervin—Federal University of Santa Catarina, Florianopolis, (Brazil)
Vera Afreixo—University of Aveiro (Portugal)
Virgilio Uarrota—Federal University of Santa Catarina, Florianopolis (Brazil)
Yinbo Cui—National University of Defense Technology (China)

## Organising Committee

Javier Bajo—Pontifical University of Salamanca (Spain)
Sara Rodríguez—University of Salamanca (Spain)
Dante I. Tapia—University of Salamanca (Spain)
Fernando de la Prieta Pintado—University of Salamanca (Spain)
Davinia Carolina Zato Domínguez—University of Salamanca (Spain)
Gabriel Villarrubia González—University of Salamanca (Spain)
Javier Prieto Tejedor—University of Salamanca (Spain)
Alejandro Hernández Iglesias—University of Salamanca (Spain)
Cristian I. Pinzón—University of Salamanca (Spain)
Rosa Cano—University of Salamanca (Spain)
Emilio S. Corchado—University of Salamanca (Spain)
Eugenio Aguirre—University of Granada (Spain)
Manuel P. Rubio—University of Salamanca (Spain)
Belén Pérez Lancho—University of Salamanca (Spain)
Angélica González Arrieta—University of Salamanca (Spain)
Vivian F. López—University of Salamanca (Spain)
Ana de Luís—University of Salamanca (Spain)
Ana B. Gil—University of Salamanca (Spain)
Mª Dolores Muñoz Vicente—University of Salamanca (Spain)
Jesús García Herrero—University Carlos III of Madrid (Spain)

# Contents

# A Preliminary Assessment of Three Strategies for the Agent-Based Modeling of Bacterial Conjugation

**Antonio Prestes García and Alfonso Rodríguez-Patón**

**Abstract** Bacterial conjugation is a cell-cell communication by which neighbor cells transmit circular DNA strands called plasmids. The transmission of these plasmids has been traditionally modeled using differential equations. Recently agent-based systems with spatial resolution have emerged as a promising tool that we use in this work to assess three different schemes for modeling the bacterial conjugation. The three schemes differ basically in which point of cell cycle the conjugation is most prone to happen. One alternative is to allow a conjugative event occurs as soon a suitable recipient is found, the second alternative is to make conjugation equally like to happen throughout the cell cycle and finally, the third one technique to assume that conjugation is more likely to occur in a specific point late in the cell cycle.

**Keywords** Agent-based modeling · Individual-based modeling · Plasmids · Bacterial conjugation · Synthetic biology

## 1 Introduction

The conjugation is a form of horizontal gene transfer where conjugative plasmids are transferred from cell to cell in a bacterial population. Plasmids may carry useful traits for their hosts and conjugation is the main cause of nosocomial antibiotic resistance. Conjugative plasmids are double stranded DNA which replicate independently from bacterial main chromosome, containing also the required genes to express the conjugative machinery, including the conjugative pili, making possible the ssDNA transfer across cellular envelopes. Infection rates are affected by the alternation periods of transitory derepression and repression cycles. Transitory derepression is deemed to last a few generations facilitating the plasmid maintenance. But even plasmids having the pilus synthesis constitutively derepressed, which represents

A.P. García · A. Rodríguez-Patón (✉)
Departamento de Inteligencia Artificial, Universidad Politécnica de Madrid, Campus de Montegancedo s/n, Boadilla Del Monte, 28660 Madrid, Spain
e-mail: arpaton@fi.upm.es



R. Overbeek et al. (eds.), *9th International Conference on Practical Applications of Computational Biology and Bioinformatics*, Advances in Intelligent Systems and Computing 375, DOI 10.1007/978-3-319-19776-0_1

the optimal conditions to undertake successful conjugative events in spatially structured colonies, are unable to completely infect if nutrients are not replenished [4]. It has been sugested that growth dependence [10, 12, 13] could explain the limited invasion in structured colonies. In this work we have built a spatially explicit [2] individual-based [7] model of bacterial conjugation which has been used to verify the suitability of three alternative methods for modeling the conjugative event. In the next sections we will describe the model and provide an early analysis of the model output.

# 2 Model Description

In this section we provide a brief description of our model using the ODD (Overview, Design concepts and Detail) protocol [5, 6]. The model was implemented completely in java language using Repast Simphony agent-based simulation framework [11].

## 2.1 *Purpose*

The objective of this model is the assessment of what strategy for modeling and implementing the rule of conjugation provides the best fit to experimental data and better captures the real structure of conjugative plasmid propagation in a bacterial population. In order to do so a very simple model was implemented and their results compared to the experimental data obtained from wet-lab. The key points of our model lie on the idea of the existence of a local or intrinsic conjugation rate that has been termed $\gamma_0$ which stands for the number of plasmid transfer events, or conjugations on a cell life-cycle basis and that the infection wave speed depends directly from the point along the cell cycle when conjugative event is triggered.

## 2.2 *State Variables and Scales*

The model comprises two entity types, namely the bacterial individuals or agents and environment. The environment contains the rate limiting amount of nutrient particles required for the cell metabolism and growth. All agents evolve in a computational domain defined by a $1000 \times 1000$ μm squared lattice divided in $10^6$ cells of $1 \times 1$ μm representing a real surface of $1 \mathrm{mm}^2$. In this model the *agents* representing bacterial cells are defined individually by two main state variables, namely the *plasmid infection state* and the $t_0$. The plasmid infection states are $Q = R, D, T$, where R means recipient bacteria (plasmid free bacteria), D a donor bacteria, and T a transconjugant bacteria (R bacteria already infected by a plasmid). The $t_0$ is the time of cell birth or the time of the last cellular division, it is employed in the estimation of agent doubling time used in the division decision rule. The T4SS pili are also taken into account and the agents have a state variable representing the number of pilus already expressed and available in cell surface.

## 2.3 Process Overview and Scheduling

The dynamics of bacterial conjugation is modeled as the execution of following set of cellular processes: the cellular division, the T4SS pili expression, the shoving relaxing which avoid bacterial cells to overlap and allow a more realistic population growth and the conjugation process. The state variable update is asynchronous. The order of execution of this process is shuffled to avoid any bias due to a purely sequential execution of model rule base. The conjugation process is modeled in three different ways with respect to the time when conjugation event is most prone to happen and the results are compared. Thus the conjugation is defined by two variables: the value of intrinsic conjugation rate ($\gamma_0$) which determines how many transfers should be performed by a single bacterial cell and the cell cycle point which defines the time when the conjugative events must occur.

## 2.4 Initialization

The simulation model is initialized with a population of plasmid free ($R$) and plasmid bearing ($D$) cells according to input parameters. The agents are placed randomly within a circular surface centered over the lattice central position. The radius of circle where agents are placed is calculated as function of $N_0$ in order to be consistent to the desired initial cell density. The simulation environment is also initialized with a number of nutrient particles in order to support the half of the estimated number of cellular divisions and the rationale behind it is to capture the intercellular competition for nutrient access.

## 2.5 Input

The model input and initialization requires the parameters shown in Table 1. The *costT4SS* is the total cost of pili expression. The cost applied for a single pilus expression is *costT4SS/param(maxpili)*. The *param(maxpili)* is actually a constant having the value of 5 for *E. coli* [1]. The *cellCycle* parameter indicates two things: the type of modeling rule and its parameter.[1]

[1] A value of −1 set the model to conjugate as soon an infected cell finds a susceptible one. Setting the parameter to 0 will randomize the conjugation time between $t_0$ and $G$. Finally using a value greater than zero indicates the specific point in the cell cycle for conjugation. A polynomic equation fitted to the experimental time series where the dependent variable is equal to $T/(T+R)$ rate.

**Table 1** The complete list of model initialization parameters

| Parameter | Unit | Description |
|---|---|---|
| G | minutes | Average doubling time for plasmid free cells |
| cellCycle | % of G | The percentage of cellular cycle for conjugation |
| costConjugation | % of G | The penalization due to a conjugative event |
| costT4SS | % of G | The Pilus expression cost |
| $\gamma_0$ | conjugations/cell | Upper limit for conjugations performed by an agent |
| isConjugative | true—false | Defines a conjugative or a mobilizable plasmid |
| isRepressed | true—false | The T4SS expression state for the plasmid |
| $N_0$ | cells/ml | Initial population expressed in cells/ml |
| donorRatio | % of $N_0$ | The initial density of donor cells ($D$) |
| Equation | N/A | An equation for experimental data |

## 2.6 Sub-models

For the sake of brevity we only mention the auxiliary process used in this model. Thus the both **Nutrient Uptake** and **Nutrient Diffusion** are based on approaches similar to those described in [9]. The **Shoving Relaxation** process was adapted from [8]. Finally the **T4SS expression** models the expression of conjugative pili which is required for conjugative events.[2] We have implemented a straightforward and simplified version of the real process but capturing yet the most significant aspects of the pili expression subsystem, namely the intra and inter-cellular competition. The intra-cellular competition is modeled as a cost in the cell doubling time applied every time the cells require the expression of conjugative apparatus. The inter-cellular competition is achieved by requiring the uptake of a nutrient particle for the expression of pili. **Conjugative transfer**—We modeled the conjugation process using three different approaches which we have called **strategies** being the distinctive point between them the decision criteria for the time when the conjugative event will be signaled. Thus, the first strategy does not take time into account and conjugation will be triggered simply when any recipient cell is found on the donor neighborhood, see Fig. 1.

The second strategy, see Fig. 2, uses a random number drawn from a uniform distribution being the number in a range of 1 and the cell doubling time, thus on average, conjugation is most prone to happen at a 50 % of cell cycle but with a high standard deviation which is given by the expression G-1/3.464102.[3]

[2]The Type IV secretion systems (T4SS) is a sort of transmembrane protein responsible, amongst other things, for the mating pair stabilization and the injection of single strand DNA into the target recipient cell.

[3]The expression comes from the common expression for estimate the standard deviation of uniform distribution which is given by $(b-a)/\sqrt(12)$ or $(b-a)/3.464102$.

Fig. 1 The Conjugation Strategy 1 algorithm

```
1: procedure CONJUGATIONSTRATEGY1
2:     if S ≥ 0 and Neighbors(R) ≥ 0 then
3:         if state(γ0) < Z(param(γ0)) then
4:             cost ← p × param(costConjugation)
5:             state(G) ← state(G) × cost
6:             Conjugate()
7:             neighbor.state(repressed) ← false
8:             state(γ0) ← state(γ0) + 1
9:         end if
10:    end if
11: end procedure
```

Fig. 2 The Conjugation Strategy 2 algorithm

```
1: procedure CONJUGATIONSTRATEGY2
2:     if S ≥ 0 and Neighbors(R) ≥ 0 then
3:         conjugationTime ← RANDOM(1, G)
4:         Δt ← t − t0
5:         if Δt ≥ conjugationTime then
6:             if state(γ0) < Z(param(γ0)) then
7:                 cost ← p × param(costConjugation)
8:                 state(G) ← state(G) × cost
9:                 Conjugate()
10:                neighbor.state(repressed) ← false
11:                state(γ0) ← state(γ0) + 1
12:            end if
13:        end if
14:    end if
15: end procedure
```

Finally in the strategy 3, see Fig. 3, a more specific time selection mechanism is used being the uniform random variable replaced by a normally distributed one with a small coefficient as low as a ten percent which is approximately three times lower than the case of strategy 2. In all strategies the actual value of local $state(\gamma_0)$ is compared to a normally distributed random variable $Z(\gamma_0)$.[4] The $state(\gamma_0)$ is just the number of conjugations which have already been performed by the agent. All strategies also check for the nutrient availability ($S$) and the presence of infectable $R$ cells on the agent neighborhood.

[4] generated using the model parameter $param(\gamma_0)$ and assuming again a coefficient of variation equals to 0.1 ($Z(\gamma_0) = 0.1\gamma_0 Z + \gamma_0$).

1: **procedure** CONJUGATIONSTRATEGY3
2: **if** $S \geq 0$ **and** $Neighbors(R) \geq 0$ **then**
3: $conjugationTime \leftarrow 0.1Z + parameter(cellCycle)$
4: $\Delta_t \leftarrow t - t0$
5: **if** $\Delta_t \geq conjugationTime$ **then**
6: **if** $state(\gamma_0) < Z(param(\gamma_0))$ **then**
7: Conjugate()
8: $neighbor.state(repressed) \leftarrow false$
9: $state(\gamma_0) \leftarrow state(\gamma_0) + 1$
10: **end if**
11: **end if**
12: **end if**
13: **end procedure**

**Fig. 3** The Conjugation Strategy 3 algorithm

# 3 Results and Discussion

Using the model described in previous section we have simulated two different plasmids, namely the pSU2007[5] and the R1 plasmid. The pSU2007 plasmid express constitutively the T4SS conjugative apparatus. On the other hand the R1 plasmid is a naturally repressed one and transconjugant cells show what is known as transitory de-repression. The experimental data used for calibrate and validate the model was obtained from [3]. The parameters used in the simulation are shown in Table 2. We have run three replications of the simulation experiments and which results are summarized in Fig. 4. The behavior of generation time is consistent with the biological assumptions regarding the penalization of plasmid harboring hosts. The model provide a lot of outputs quantifying several indicators, including the estimated generation time, the percentage of infected cells due to horizontal infection (data not shown), etc.

As can be observed in Fig. 4, by a simple visual assessment, under the same initial conditions the strategy 1 and 2 overshoots the experimental curve significantly both in slope as in the final values of conjugation rates. In the case of repressed plasmid pR1 the first two strategies are very distant from the experimental. On the other hand the strategy three seems to provide naturally a good fit to the real data, being pretty close the curves and with similar shapes. As can be observed, the simulated values of pR1 slightly overestimate the experimental data and this can be attributed to a higher cost of plasmid maintenance but we have to investigate further. The strategy 1 is apparently the worst implementing the local conjugation rule for agent-based models. The second strategy, although the conjugations events occur on average at a 50 % of cell cycle is also very far from the experimental data. That seems to indicate that there are some timely mechanisms which control the plasmid infection wave in a bacterial colony. The exact underlying mechanism is not known but our simulation

[5]This plasmid is based on the R388 backbone.

**Table 2** The initialization parameters for conjugation experiments. The first column is the name of real plasmid being simulated and the number inside parentheses indicates what strategy is being used. The column *Conjugation* is corresponds to *param(costConjugation)*

| Name | G | cellCycle | Conjugation | costT4SS | $\gamma_0$ | $N_0$ | donorRatio | Repressed |
|---|---|---|---|---|---|---|---|---|
| pSU2007(S1) | 43 | N/A (-1) | 5 % | 60 % | 3 | $9 \times 10^9$ | 50 % | False |
| pSU2007(S2) | 43 | N/A (0) | 5 % | 60 % | 3 | $9 \times 10^9$ | 50 % | False |
| pSU2007(S3) | 43 | 75 % | 5 % | 60 % | 3 | $9 \times 10^9$ | 50 % | False |
| R1(S1) | 43 | N/A (-1) | 5 % | 40 % | 2 | $9 \times 10^9$ | 50 % | True |
| R1(S2) | 43 | N/A (0) | 5 % | 40 % | 2 | $9 \times 10^9$ | 50 % | True |
| R1(S3) | 43 | 75 % | 5 % | 40 % | 2 | $9 \times 10^9$ | 50 % | True |

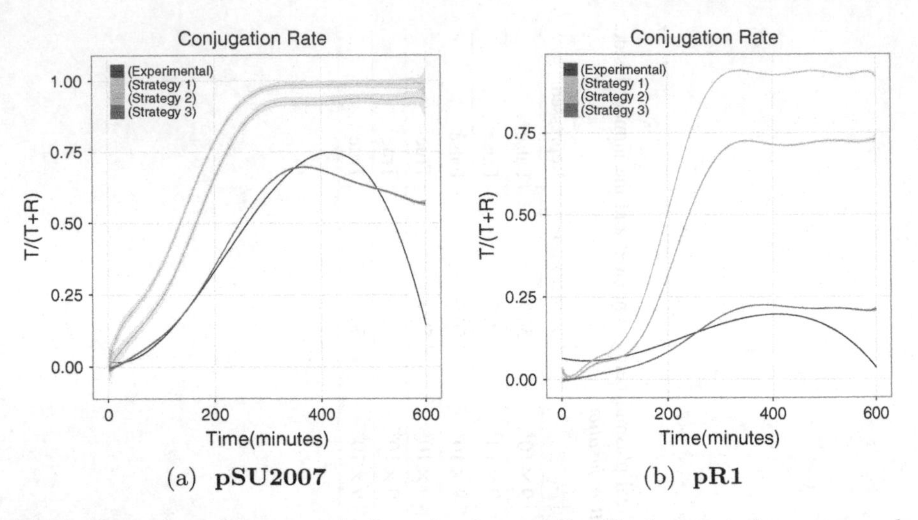


(a) **pSU2007** (b) **pR1**

**Fig. 4** Comparison of three strategies for conjugation modeling. The three strategies are compared to the experimental (black line) data (Color figure online)

results point that if conjugation is allowed to happen too early in the cellular cycle the plasmid is flooded to the whole colony at a much higher speed that would be expected.

Another interesting and apparently counter-intuitive aspect is the effect of conjugation pace in the average values of $\gamma_0$, as can be seen in Fig. 5. The first two strategies despite of the higher speed of infection wave have lower values of $\gamma_0$ than the strategy 3. The cause of that apparent contradiction can be attributed to the fact that if cells are able to forward the plasmid earlier more cells are infected and able to infect soon which implies an average small number of conjugations performed on a cell basis. It is also worth noting that values of $\gamma_0$ estimated by all strategies are, on average, lower than the value used as the model parameter. This is due to the effect of other constraints existing in model, namely the neighborhood structure or nutrient availability which imposes an upper limit for the values of $\gamma_0$.

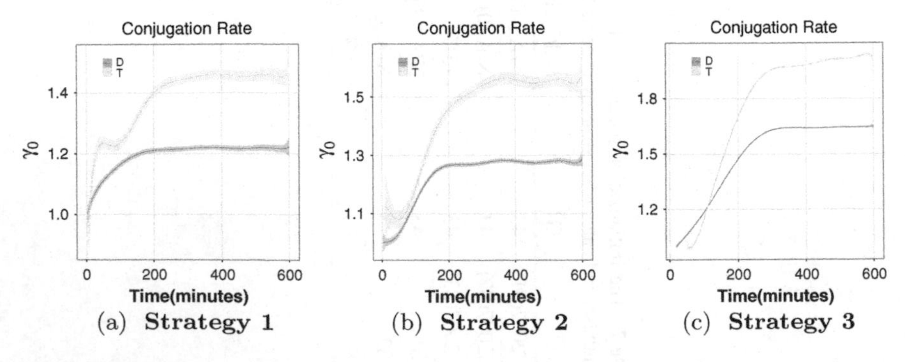


(a) **Strategy 1** (b) **Strategy 2** (c) **Strategy 3**

**Fig. 5** $\gamma_0$ values for pSU2007 strategies. Yellow curve (Donors), red curve (Transconjugants) (Color figure online)

**Acknowledgments** This work was supported by the European FP7-ICT-FET EU research project 612146 (PLASWIRES "Plasmids as Wires" project) www.plaswires.eu and by Spanish Government (MINECO) research grant TIN2012-36992.

## References

1. Arutyunov, D., Frost, L.S.: F conjugation: back to the beginning. Plasmid **70**(1), 18–32 (Jul 2013)
2. Berec, L.: Techniques of spatially explicit individual-based models:construction, simulation, and mean-field analysis. Ecol. Model. **150**(1–2), 55–81 (April 2002)
3. del Campo, I., Ruiz, R., Cuevas, A., Revilla, C., Vielva, L., de la Cruz, F.: Determination of conjugation rates on solid surfaces. Plasmid **67**(2), 174–182 (March 2012)
4. Fox, R.E., Zhong, X., Krone, S.M., Top, E.M.: Spatial structure and nutrients promote invasion of IncP-1 plasmids in bacterial populations. ISME J. **2**(10), 1024–1039 (June 2008)
5. Grimm, V., Berger, U., Bastiansen, F., Eliassen, S., Ginot, V., Giske, J., Goss-Custard, J., Grand, T., Heinz, S.K., Huse, G., Huth, A., Jepsen, J.U., Jørgensen, C., Mooij, W.M., Müller, B., Pe'er, G., Piou, C., Railsback, S.F., Robbins, A.M., Robbins, M.M., Rossmanith, E., Rüger, N., Strand, E., Souissi, S., Stillman, R.A., Vabø R., Visser, U., DeAngelis, D.L.: A standard protocol for describing individual-based and agent-based models. Ecol. Model. **198**(1–2), 115–126 (Sept 2006)
6. Grimm, V., Berger, U., DeAngelis, D.L., Polhill, J.G., Giske, J., Railsback, S.F.: The ODD protocol: a review and first update. Ecol. Model. **221**(23), 2760–2768 (Nov 2010)
7. Hellweger, F.L., Bucci, V.: A bunch of tiny individuals—individual-based modeling for microbes. Ecol. Model. **220**(1), 8–22 (Jan 2009)
8. Kreft, J.U., Booth, G., Wimpenny, J.W.T.: BacSim, a simulator for individual-based modelling of bacterial colony growth. Microbiology **144**(12), 3275–3287 (Dec 1998)
9. Krone, S.M., Lu, R., Fox, R., Suzuki, H., Top, E.M.: Modelling the spatial dynamics of plasmid transfer and persistence. Microbiology (Reading, England) **153**(Pt 8), 2803–2816 (Aug 2007)
10. Merkey, B.V., Lardon, L.A., Seoane, J.M., Kreft, J.U., Smets, B.F.: Growth dependence of conjugation explains limited plasmid invasion in biofilms: an individual-based modelling study. Environ. Microbiol. **13**(9), 2435–2452 (2011)
11. North, M., Collier, N., Ozik, J., Tatara, E., Macal, C., Bragen, M., Sydelko, P.: Complex adaptive systems modeling with Repast Simphony. Complex Adapt. Syst. Model. **1**(1), 1–26 (2013)
12. Simonsen, L., Gordon, D.M., Stewart, F.M., Levin, B.R.: Estimating the rate of plasmid transfer: an end-point method. J. Gen. Microbiol. **136**(11), 2319–2325 (Nov 1990)
13. Zhong, X., Droesch, J., Fox, R., Top, E.M., Krone, S.M.: On the meaning and estimation of plasmid transfer rates for surface-associated and well-mixed bacterial populations. J. Theor. Biol. **294**, 144–152 (Feb 2012)

# Carotenoid Analysis of Cassava Genotypes Roots (*Manihot Esculenta* Crantz) Cultivated in Southern Brazil Using Chemometric Tools

**Rodolfo Moresco, Virgílio G. Uarrota, Aline Pereira, Maíra Tomazzoli, Eduardo da C. Nunes, Luiz Augusto Martins Peruch, Christopher Costa, Miguel Rocha and Marcelo Maraschin**

**Abstract** *Manihot esculenta* roots rich in β-carotene are an important staple food for populations with risk of vitamin A deficiency. Cassava genotypes with high provitamin A activity have been identified as a strategy to reduce the prevalence of deficiency of this vitamin, In this study, the metabolomics characterization focusing on the carotenoid composition of ten cassava genotypes cultivated in southern Brazil by UV-visible scanning spectrophotometry and reverse phase-high performance liquid chromatography was performed. The data set was used for the construction of a descriptive model by chemometric analysis. The genotypes of yellow roots were clustered by the higher concentrations of *cis*-β-carotene and lutein. Inversely, cream roots genotypes were grouped precisely due to their lower concentrations of these pigments, as samples rich in lycopene differed among the studied genotypes. The analytical approach (UV-Vis, HPLC, and chemometrics) used showed to be efficient for understanding the chemodiversity of cassava genotypes, allowing to classify them according to important features for human health and nutrition.

**Keywords** Chemometrics · Descriptive models · Partial metabolome · Cassava genotypes · Carotenoids · RP-HPLC · UV-vis

R. Moresco (✉) · V.G. Uarrota · A. Pereira · M. Tomazzoli · M. Maraschin
Plant Morphogenesis and Biochemistry Laboratory, Federal University of Santa Catarina, Florianopolis, Brazil
e-mail: rodolfo.moresco@posgrad.ufsc.br

E. da C. Nunes · L.A.M. Peruch
Santa Catarina State Agricultural Research and Rural Extension Agency (EPAGRI), Experimental Station of Urussanga, Urussanga, Brazil

C. Costa · M. Rocha
Centre Biological Engineering School of Engineering, University of Minho, Braga, Portugal



R. Overbeek et al. (eds.), *9th International Conference on Practical Applications of Computational Biology and Bioinformatics*, Advances in Intelligent Systems and Computing 375, DOI 10.1007/978-3-319-19776-0_2

# 1 Introduction

Cassava (*Manihot esculenta* Crantz, 1766) currently ranks as the third most important species as a source of calories in the world among the group of staple food crops, including rice and maize. It is primarily consumed in places where there are prevailing drought, poverty, and malnutrition [1]. Diseases related to vitamin A deficiency are among the major nutritional problems in developing countries. It is estimated that 190 million children in preschool age have low retinol activity in plasma (<0.70 μmol.L$^{-1}$), subclinical symptom of this deficiency [2]. Cassava is considered a staple food for many populations with risk of vitamin A deficiency and is predominantly produced by small-scale farmers with limited resources. In cassava, β-carotene is the majoritarian pro-vitamin A carotenoid [3], but the amounts found in white and cream cassava roots (the most commonly consumed by populations) are low [4]. It is known that roots with yellow flesh are highly correlated with the concentration of carotenoids and the search for materials with higher pro-vitamin A activity is recognized as a strategy to reduce the prevalence of this deficiency [5]. Besides β-carotene, *M. esculenta* also contains small amounts of other carotenoids, e.g., lycopene and the xanthophylls, lutein and β-cryptoxanthin, with recognized health benefits [6].

Because of the high importance of cassava crops in Brazil, genebanks of cassava collections have been established and maintained for the purpose of preserving commercial varieties, traditional landraces and wild genotypes. Thus, the identification and preservation of genotypes with high carotenoid concentrations is thought to be relevant for the Brazilian and global food security and nutrition.

This study, in connection with the metabolomics characterization of the genebank's cassava accesses, emphasizes their carotenoid profile in root samples, using a typical analytical platform, i.e., UV-visible spectrophotometry (UV-Vis) and reverse phase-high performance liquid chromatography (RP-HPLC). The data set afforded (i.e., the spectrophotometric profiles and the chromatographic quantification of each carotenoid compound) was used to build descriptive and classification models by calculation of the principal components and cluster analysis. Such an analytical approach allows the rapid and effective extraction of relevant and non-redundant information from a set of complex data, enabling a more detailed and robust understanding of possible differences and/or similarities in the studied samples, as well as their improved discrimination. In practical terms, this study develops and applies biotechnological approaches, by coupling the use of biochemical markers with bioinformatics tools in order to gain insights to support genetic breeding programs of cassava.

# 2 Materials and Methods

Roots of ten genotypes of *M. esculenta* (2010/2011 season) from the EPAGRI's germplasm bank (Urussanga Experimental Station, 28º31'18''S, 49º19'03''W, Santa Catarina, southern Brazil) were used in this study. Traditionally, they have been called

*Apronta mesa, Pioneira, Oriental, Amarela, Catarina, IAC, Salézio, Estação, Videira* and *Rosada* and were selected based on their economic and social importance.
Carotenoids were extracted from fresh roots as described by Rodriguez-Amaya & Kimura (2004) [7] using an Ultra-Turrax (Janke & Kunkel IKA - T25 basic) and organic solvents: acetone and petroleum ether. The absorbances of the organosolvent extracts were collected on a UV-visible spectrophotometer (Gold Spectrum lab 53 UV-Vis spectrophotometer, BEL photonics, Brazil) using a spectral window from 200 to 700 nm. Aliquots (10 µl) of the extracts were also injected into a liquid chromatograph (LC-10A Shimadzu) system equipped with a C18 reversed-phase column (Vydac 201TP54, 250 mm × 4.6 mm, 5 µm Ø, 35 °C) coupled to a pre-column (C18 Vydac 201TP54, 30 mm × 4.6 mm, 5 µm Ø) and a spectrophotometric detector (450 nm). Methanol: acetonitrile (90: 10, v/v) was used for elution at a rate of 1 ml/min. The identification and quantification of compounds of interest was carried out via co-chromatography and comparison of retention times of samples with those of standard compounds (Sigma–Aldrich, USA) under the same experimental conditions.
Data were collected, summarized, and submitted to analysis of variance (ANOVA) followed by post hoc Tukey's test ($p < 0.05$) for mean comparison. All procedures were performed in triplicate ($n = 3$). The processing of the spectrophotometric profile considered the definition of the spectral window of interest (200–700 nm), baseline correction, normalization, and optimization of the signal/noise ratio (smoothing). The processed data set was initially subjected to multivariate statistical analysis, by applying principal component analysis (PCA) and clustering methods. Furthermore, the spectral data set and the amounts of the target carotenoids determined by RP-HPLC were used for calculation of the principal components, supported by scripts written in R language (v. 3.1.1) [8] using tools defined by our research group and some functions from the packages Chemospec [9], HyperSpec [10], and ggplot2. All R scripts, raw data, and additional chemometric analysis are available in supplementary material, in http://darwin.di.uminho.pt/metabolomicspackage/ as well as the data analysis report automatically generated from the R scripts using the features provided by R Markdown (http://darwin.di.uminho.pt/metabolomicspackage/cassava-carotenoids.html). This allows anyone to fully reproduce and document the experiments.

## 3 Results and Discussion

Carotenoids show typically maximum absorption at 450 nm [7] and as depicted in Fig. 1, all the spectral profiles (200–700 nm) of the yellow and red cassava root extracts revealed prominent absorbance peaks between 400 - 500 nm, indicating that the organosolvent system used was efficient to extract the target metabolites. Lower absorbance values were found in cream color-roots, precisely because they have low concentrations of carotenoids as the *Rosada* genotype (pink root) showed the highest absorbance values at 450 nm.

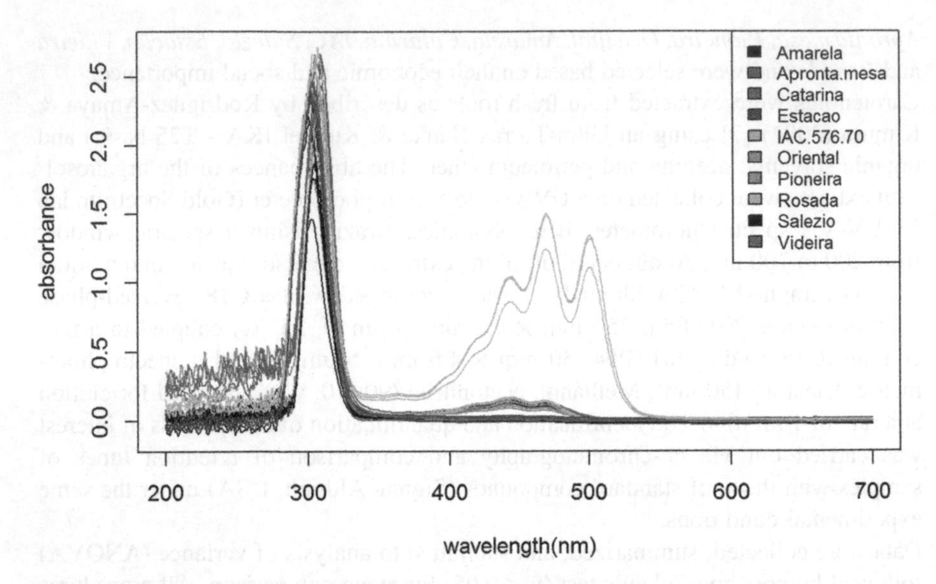


**Fig. 1** Typical UV-Vis spectrophotometric profiles (λ = 200–700 nm, acetone: petroleum ether - v/v) of root parenchymal tissues of ten cassava genotypes cultivated in southern Brazil

When the principal components were calculated from the full spectrophotometric profile (λ = 200–700 nm) data matrix, PC1 and PC2 contributed to explain 78.9 % of the total variance of the data set. However, a clear discrimination of the samples according to the carotenoid concentrations was not found. Only the *Rosada* genotype distinguished from the others by grouping in PC1 + / PC2 -. Genotypes with high (yellow roots) and low carotenoid contents (cream roots) were spread out over the factorial distribution plane, making difficult to discriminate between them (Fig. 2).

Such findings prompted us to build a second analytical model by applying PCA to the carotenoid fingerprint region of the UV-visible (400–500 nm). In this case, PC1 and PC2 accounted for 99.97 % of the variance, clearly revealing three groups according to their similarities (Fig. 3). Interestingly, the samples were grouped according to their carotenoids contents determined by RP-HPLC and distributed according to the root flesh color. Cassava genotypes with yellow roots (*Pioneira, Amarela, Catarina* and *IAC576-70*) were clustered along the PC2 + axis. Genotypes with cream roots and lower carotenoid content (*Apronta mesa, Oriental, Salézio, Estação* and *Videira*) were grouped in PC1/PC2 −. In its turn, *Rosada* genotype (flesh red) seems to have a metabolic profile occurring away from all the other samples.

The chromatographic profiles of the organosolvent extracts identified *cis*- and *trans*-β-carotene, α-carotene, lutein, and β-cryptoxanthin in all the cassava genotypes analyzed. The presence of lycopene, a common precursor of the carotenoids above mentioned, was detected only in *Rosada* genotype, a fact that led us to speculate this is an important reason for its clear discrimination in respect to other genotypes.

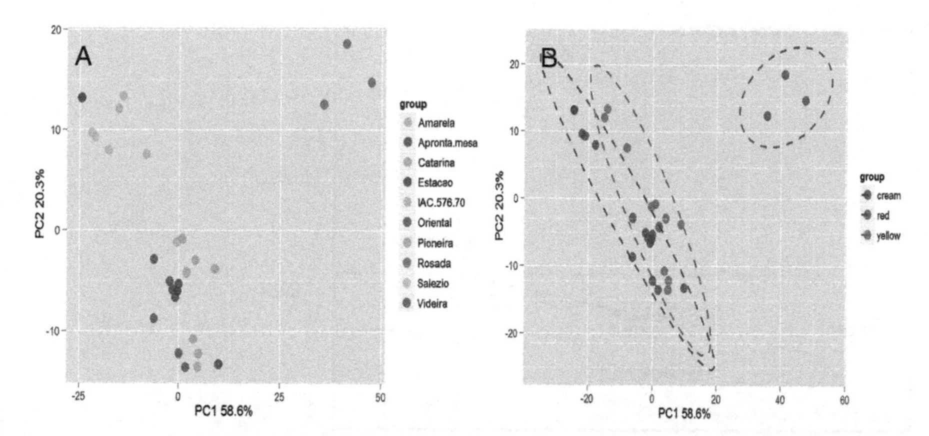


**Fig. 2** A - Factorial distribution (principal components 1 and 2) of the spectral data set (UV-Vis, 200–700 nm) of the organosolvent extract of roots of ten cassava genotypes. B - Graphical demonstration according to the root flesh color

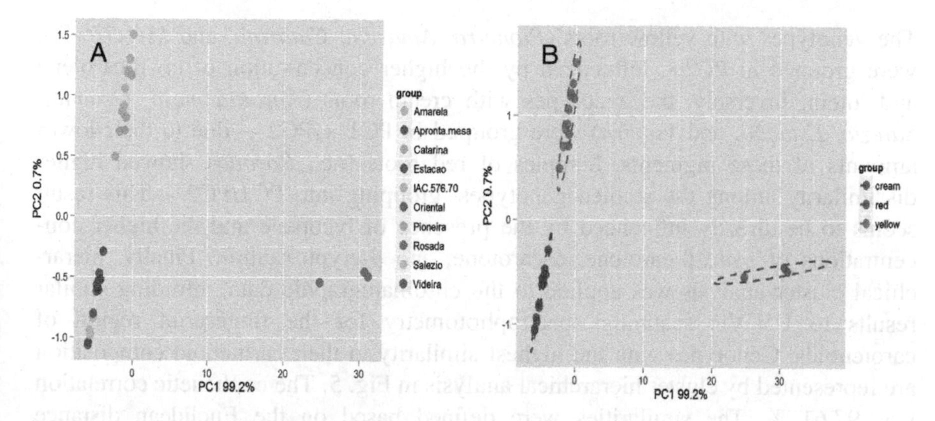


**Fig. 3** A - Factorial distribution (principal components 1 and 2) of the spectral data set of the fingerprint region of carotenoids (UV-Vis 400–500 nm, acetone: petroleum ether - v/v). B - Graphical demonstration according to the root flesh color

The isomer *trans* of β-carotene was the major compound regardless the sample analyzed.

In a second series of experiments, PCA was applied to the chromatographic data set revealing patterns of similarity of carotenoid composition among the studied genotypes. These findings corroborate the results previously found by UV-Vis scanning spectrophotometry taking into account the fingerprint region of carotenoids (i.e. 400–500 nm). Figure 4 depicts the grouping of genotypes after calculation of the principal components from the RP-HPLC quantification of carotenoids. PC1 and PC2 explain 97.8 % of the total variance of the sample population under study.

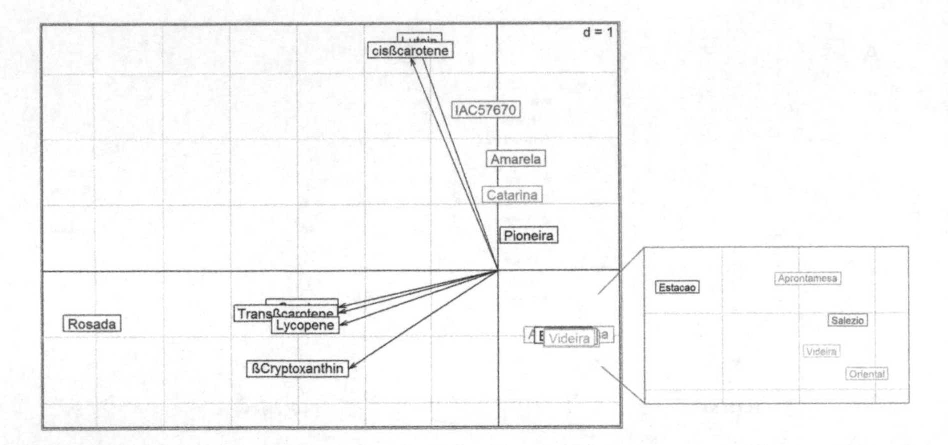


**Fig. 4** A - Score-scatter plot (PC1 and PC2) of the quantitative data of carotenoids determined by RP-HPLC in root samples of ten cassava genotypes (n = 3). B - Magnification to the overlapping samples at the PCA

The genotypes with yellow roots (*Pioneira, Amarela, Catarina*, and *IAC-576-70*) were grouped in PC2+, influenced by the higher concentration of *cis*-β-carotene and lutein. Inversely, the genotypes with cream roots (*Apronta mesa, Oriental, Salézio, Estação*, and *Videira*) were grouped in PC1 +/PC2 – due to their lower amounts of these pigments. Samples of red roots (i.e., *Rosada*) showed higher dissimilarity among the studied genotypes, grouping into PC1/PC2 -. This result seems to be directly influenced by the presence of lycopene and the higher concentrations of *trans*-β-carotene, α-carotene, and β-cryptoxanthin. Finally, hierarchical cluster analysis was applied to the chromatographic data, affording similar results to UV-Vis scanning spectrophotometry for the fingerprint region of carotenoids. Genotypes with the highest similarity in their carotenoid composition are represented by cluster hierarchical analysis in Fig. 5. The cophenetic correlation was 97.61 %. The similarities were defined based on the Euclidean distance between two samples using the arithmetic average (UPGMA).

## 4 Conclusions

The data set obtained by the analytical techniques employed in this work allowed a better understanding of the chemical variability associated with roots' carotenoid composition of the cassava genotypes. The large disparity in carotenoid contents reveals that there is a chemical variability between genotypes analyzed. The cassava genotypes showed substantial amounts of carotenoids, indicating their potential as source of interesting compounds to human health and nutrition, given the presence of pro-vitamin A carotenoids (β-carotene, e.g.) and lycopene in roots of yellow and red color, respectively. The *Rosada* genotype was found to be discrepant because

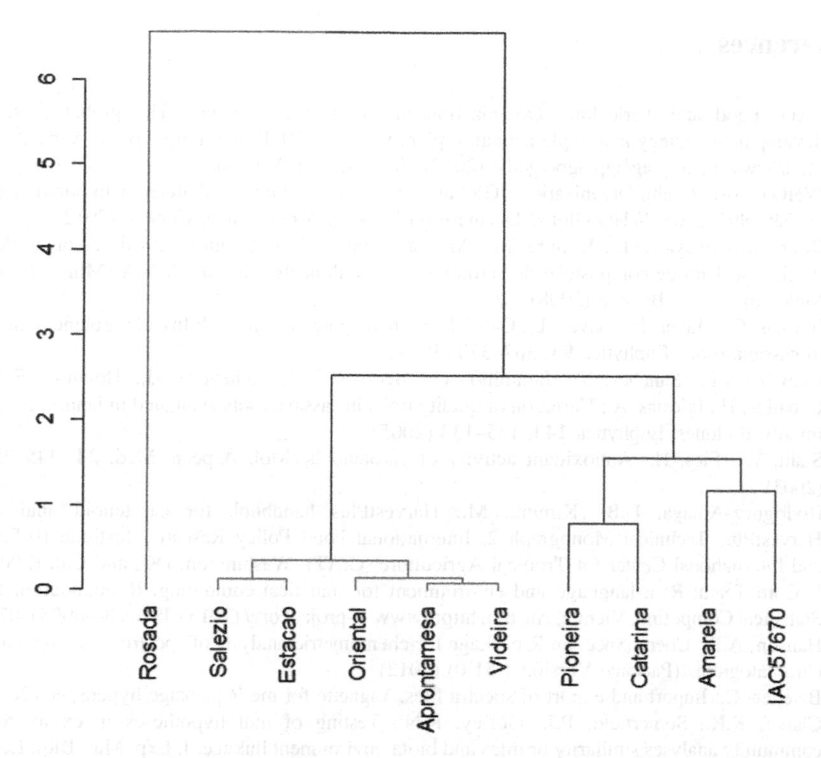


**Fig. 5** Similarity of cassava genotypes in respect to their carotenoid composition determined by RP-HPLC, followed by hierarchical clustering analysis (UPGMA method - 97.61 % of cophenetic correlation). The genotypes similarity between members of the same cluster is statistically significant ($p < 0.05$) when the branches in the dendrogram show the same color.[1]

its richness in the carotenoids, in addition to the presence of lycopene in relevant amounts.

The information obtained by coupling the analysis of biochemical markers for pro-vitamin A in cassava genotypes to bioinformatics tools revealed to be relevant for the rational design of biochemically-assisted cassava breeding programs. Indeed, the analytical approach adopted (i.e., UV-Vis/RP-HPLC/chemometrics) allowed to discriminate and classify the claimed genetic variability of the studied samples based on their biochemical traits, helping to identify/select cassava genotypes of interest to human health and nutrition.

**Acknowledgements** To FAPESC (Fundação de Amparo à Pesquisa e Inovação do Estado de Santa Catarina) and CNPq (Conselho Nacional de Desenvolvimento Científico e Tecnológico) for financial support. The research fellowship from CNPq on behalf of the later author is acknowledged.

---

[1]Significance determined by Simprof analysis (Similarity Profile Analysis) from R Clustsig package in accordance with Clarke, Somerfield & Gorley (2008) [11].

## References

1. FAO (Food and Agriculture Organization of the United Nations): The global cassava development strategy and implementation plan, vol. 1, p. 70. Rome, Italy, 26–28 April 2000. http://www.fao.org/ag/agp/agpc/gcds/ (2012). Accessed 13 Apr 2012
2. WHO (World Health Organization): Global prevalence of vitamin A deficiency in populations at risk 1995–2005. WHO Global Database on Vitamin A deficiency, Geneva (2012)
3. Rodrigues-Amaya, D.B., Kimura, M., Amaya-Farfan, J.: Fontes brasileiras de carotenoides: tabela brasileira de composição de carotenoides em alimentos, p. 100. MMA (Ministério do Meio Ambiente), Brasília (2008)
4. Iglesias, C., Mayer, J., Chavez, L., Calle, F.: Genetic potential and stability of carotene content in cassava roots. Euphytica **94**, 367–373 (1997)
5. Chavéz, A.L., Sánchez, T., Jaramillo, G., Bedoya, J.M., Echeverry, J., Bolaños, E.A., Ceballos, H., Iglesias, A.: Variation of quality traits in cassava roots evaluated in landraces and improved clones. Euphytica **143**, 125–133 (2005)
6. Stahl, W., Sies, H.: Antioxidant activity of carotenoids. Mol. Aspects Med. **24**, 345–351 (2003)
7. Rodriguez-Amaya, D.B., Kimura, M.: HarvestPlus handbook for carotenoid analysis. HarvestPlus Technical Monograph 2. International Food Policy Research Institute (IFPRI) and International Center for Tropical Agriculture (CIAT), Washington, DC, and Cali (2004)
8. R Core Team: R: a language and environment for statistical computing. R Foundation for Statistical Computing, Vienna, Austria. http://www.R-project.org/ (2015). ISBN 3-900051-07-0
9. Hanson, A.B.: ChemoSpec: an R package for chemometric analysis of spectroscopic data and chromatograms (Package Version 1.51-0) (2012)
10. Beleites, C.: Import and export of spectra files. Vignette for the R package hyperSpec (2011)
11. Clarke, K.R., Somerfield, P.J., Gorley, R.N.: Testing of null hypotheses in exploratory community analyses similarity profiles and biota-environment linkage. J. Exp. Mar. Biol. Ecol. **366**, 56–69 (2008)

# UV-Visible Scanning Spectrophotometry and Chemometric Analysis as Tools to Build Descriptive and Classification Models for Propolis from Southern Brazil

**Maíra M. Tomazzoli, Remi D. Pai Neto, Rodolfo Moresco, Larissa Westphal, Amélia R.S. Zeggio, Leandro Specht, Christopher Costa, Miguel Rocha and Marcelo Maraschin**

**Abstract** Propolis is a chemically complex biomass produced by honeybees (*Apis mellifera*) from plant resins added of salivary enzymes, beeswax, and pollen. Recent studies classified Brazilian propolis into 12 groups based on physiochemical characteristics and different botanical origins. Since propolis quality depends, among other variables, on the local flora which is strongly influenced by (a)biotic factors over the seasons, to unravel the harvest season effect on the propolis' chemical profile is an issue of recognized importance. For that, fast, cheap, and robust analytical techniques seem to be the best choice for large scale quality control processes in the most demanding markets, e.g., human health applications. UV-Visible (UV-Vis) scanning spectrophotometry meets those prerequisites and was adopted, affording a spectral dataset containing the chemical profiles of hydroalcoholic extracts of sixty five propolis samples collected over the distinct seasons of year 2014, in southern Brazil. Descriptive and classification models were built following a chemometric approach, i.e. principal component analysis (PCA) and hierarchical clustering analysis (HCA), by using bioinformatics tools supported by scripts written in the R language. The spectrophotometric profile approach associated with chemometric analyses allowed identifying a different pattern in samples of propolis produced during the summer season over the other seasons.

M.M. Tomazzoli (✉) · R.D. Pai Neto · R. Moresco · L. Westphal
A.R.S. Zeggio · M. Maraschin
Plant Morphogenesis and Biochemistry Laboratory, Federal University of Santa Catarina, Florianopolis, Brazil
e-mail: mairatomazzoli@gmail.com

L. Specht
Environmental Military Police, Florianopolis, Brazil

C. Costa · M. Rocha
Centre Biological Engineering School of Engineering, University of Minho, Braga, Portugal



R. Overbeek et al. (eds.), *9th International Conference on Practical Applications of Computational Biology and Bioinformatics*, Advances in Intelligent Systems and Computing 375, DOI 10.1007/978-3-319-19776-0_3

Importantly, the discrimination based on PCA could be improved by using the dataset of the fingerprint region of phenolic compounds (λ = 280–350 ηm), suggesting that besides the biological activities presented by those secondary metabolites, they are also relevant for the discrimination and classification of that complex matrix through bioinformatics tools.

**Keywords** Propolis • UV-Vis scanning spectrophotometry • Chemometrics • Metabolic profile • Botanical source • Seasonality

## 1 Introduction

Propolis is a resinous substance collected by honeybees *Apis mellifera* from various plant sources and added to salivary enzymes, beeswax, and pollen. Bees use propolis to seal openings in their honeycombs and to protect them against microorganisms, and insects. Many studies have reported a broad spectrum of propolis' biological activities, e.g., cytotoxic, antiherpes, free radical scavenging, antimicrobial, and anti-HIV activities [1–4]. More recently, it has been proposed a classification system where Brazilian propolis fit into 12 groups based on their physiochemical traits and botanical origins [5]. Botanical origin of propolis is extremely important in order to guarantee raw materials of superior quality to supply demanding markets as cosmetics and pharmaceutical drugs. Previous studies of our research group (unpublished data) have identified a series of compounds in propolis produced in highland areas (i.e., São Joaquim County - altitude 1,360 m) in southern Brazil, which could be hypothetically associated to the native flora [6], giving rise to a typical propolis chemotype. In this study, a bioinformatics approach was used, applying multivariate statistical techniques (principal component analysis - PCA and hierarchical clustering analysis - HCA) to a UV-Visible scanning spectrophotometric dataset (n = 65 samples, λ = 280–800 ηm) of propolis hydroalcoholic extracts. Such analytical strategy aimed to gain insights as to the claimed chemical heterogeneity of propolis samples collected over the seasons (summer, autumn, winter, and spring) determined by the changes in the flora of the geographic region in study. Currently the development of descriptive and classification models based on fast, cheap, and robust analytical techniques such as UV-Vis is of interest to the pharmaceutical industry, for instance, since more detailed techniques (liquid or gas chromatography, coupled or not to mass spectrometry detectors) present important constraints as routine analysis for quality control of complex matrices like propolis. On the other hand, the large amounts of data afforded by UV-Vis scanning spectrophotometry and the eventual similarity of the spectral profiles of the samples turns the adoption of bioinformatics tools compulsory to obtain relevant and additional information.

# 2 Materials and Methods

## 2.1 Propolis Samples and Selection

Propolis samples from *A. mellifera* (n = 65) were collected in São Joaquim County (28°17′38″S, 49°55′54″W, Santa Catarina state, southern Brazil) during 2014, in all seasons: summer, spring, autumn, and winter. The samples were classified by visual analysis according to their colors, i.e., red, green, brown, and light brown taking into account that the resins collected by bees present a color peculiar to the donor plant.

## 2.2 Propolis Extraction and UV-Visible Scanning Spectrophotometry

The preparation of hydroalcoholic extracts was performed as described by Popova et al. (2004), with modifications [7]. Propolis samples (500 mg) were added of 25 mL 70 % ethanol (v/v) and incubated (24 h, darkness). The extracts were filtered on cellulose support under vacuum, completing the final volume to 25 mL with 70 % EtOH. The absorbances of the hydroalcoholic extracts were measured on a UV-visible spectrophotometer (Gold Spectrum lab 53 UV-Vis spectrophotometer, BEL photonics, Brazil) using a spectral window of 280 to 800 ηm (2 ηm resolution/ data point).

## 2.3 Statistical and Chemometric Analysis

The UV-Vis data set of the propolis hydroalcoholic extract was processed considering the definition of the spectral window of interest (280–800 ηm), baseline correction, normalization, and optimization of the signal/noise ratio (smoothing). Further, the data matrix was exported to Excel® datasheet as a.*csv* format file and subjected to multivariate statistical analysis, using principal component analysis (PCA) and hierarchical clustering analysis (HCA). For that, scripts were written in R language (v. 3.1.1) using tools defined by our research group and some functions from the packages Chemospec [8] and HyperSpec [9]. The scripts, raw data, and chemometric analysis are available in supplementary material in the site: http.//darwin.di.uminho.pt/metabolomicspackage. The report of analysis generated from the scripts provided by the R Markdown is available in http://darwin.di.uminho.pt/metabolomicspackage/propolis-sj.html. PCA and HCA can help one to extract relevant features from a given dataset, minimizing the redundant information and characterizing the relationship between the variables studied.

# 3 Results and Discussion

Propolis is not used as a raw material directly in industry; rather it is preprocessed by removing inert material, wax, and dirt and insoluble material, followed by the extraction of its bioactive compounds with suitable solvents. This process must preserve bioactive compounds, particularly phenolic ones. Phenolics show typically UV absorption at 290–380 ηm [10] and all the spectral profiles (280–800 ηm) of the studied samples showed signs of absorbance in that spectral window (Fig. 1), indicating that the extraction system EtOH: water (70: 30, v/v) was able to recover the phenolic compounds from propolis. Besides, the spectral profiles showed to be somewhat similar, suggesting a homogenous chemical composition among the samples, despite their collecting season. Thus, the UV-Vis spectral dataset was used for calculation of the principal components and for hierarchical clustering analysis, in order to tentatively classify the propolis samples into homogeneous groups according to the harvest season.

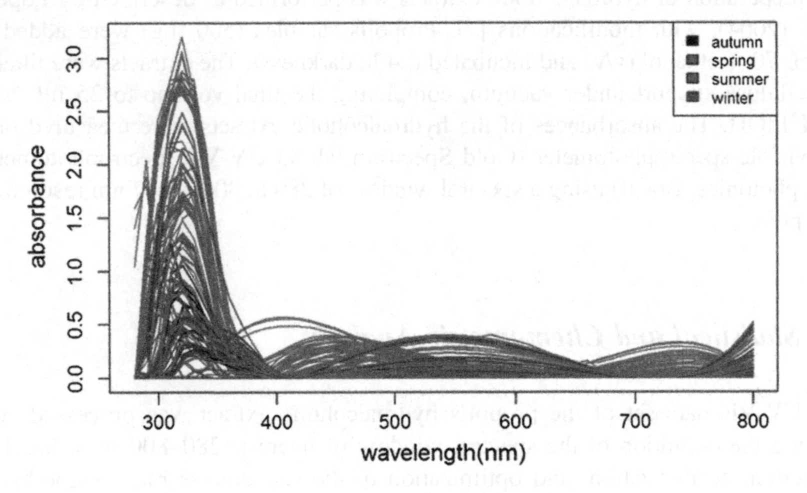


**Fig. 1** UV-Vis spectral profile (λ = 280–800 nm) of hydroalcoholic extracts (70 % EtOH) of propolis samples from southern Brazil (São Joaquim County, Santa Catarina state)

Hierarchical clustering analysis (HCA) was applied to the UV-Vis dataset (λ = 280–800 ηm - Fig. 2). The Euclidean distance between two samples using the arithmetic average (UPGMA) was used to determine the similarities. The objects in each cluster tend to be similar, but different of objects in other clusters, with no initial information on group composition [11]. It is possible from the tree to classify the samples into two main groups, the first one having samples collected in the four seasons, but with few samples collected in the summer. The second group, however, contains almost exclusively propolis samples produced in the summer, revealing an interesting separation.

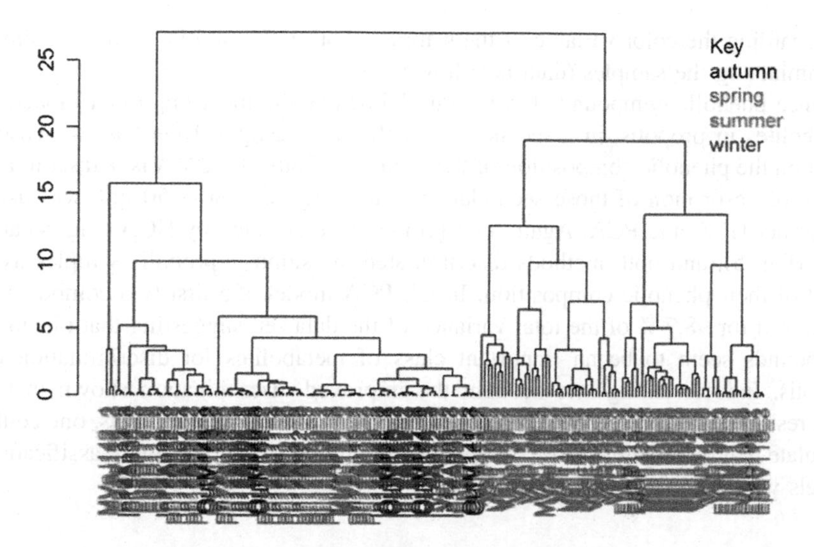


**Fig. 2** Hierarchical clustering dendogram (UPGMA method) of sixty five propolis samples from southern Brazil, according to their harvest seasons

In order to get a better understanding of the harvest season effect indicated by HCA, the UV-Vis dataset was used for the calculation of the principal components. The first two components PC1 (67.2 %) and PC2 (13.5 %) explained 80.7 % of the total variance of the dataset (Fig. 3). By expanding the model and including the contribution of the PC3 (12.1 %), it was possible to cover 92.8 % of dataset's variability. The PCA results have confirmed the sample discrimination by seasons into two groups, as observed in the HCA. The summer samples dispersed in the two components, while the remaining ones overlaid and centered on the graphic.

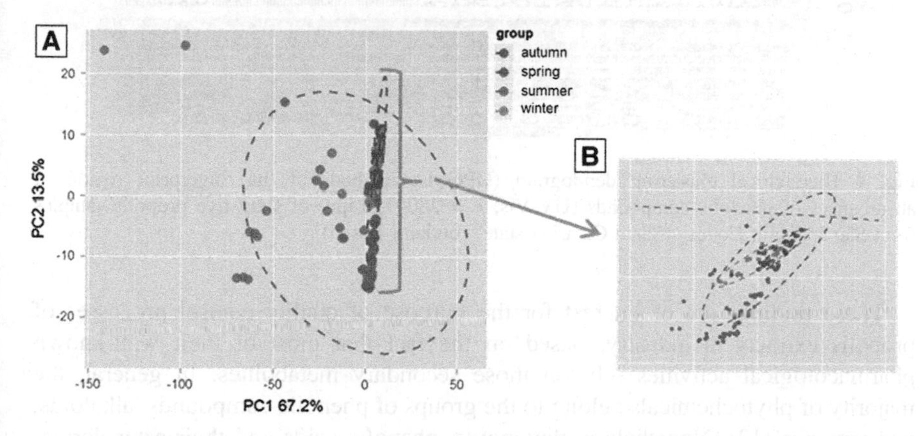


**Fig. 3** A - Principal components analysis (PCAs) scores scatter plot of the UV-Vis spectral profile ($\lambda = 280$–$800$ ηm) of sixty five propolis samples. The samples were collected over the seasons in 2014 in São Joaquim County, southern Brazil. B - Amplification to the overlapping samples at the PCA

Regarding the color variable of the samples, both HCA and PCA did not allow discriminating the samples (data not shown).

Since phenolic compounds have been claimed as the most important bioactive metabolites in propolis, in a second approach we investigated the harvest season effect on the phenolic composition of that biomass. Thus, the UV-Vis dataset in the region of absorption of those secondary metabolites, i.e., 280–350 ηm, was used for further HCA and PCA. Again, two groups were detected by HCA (Fig. 4) and PCA (Fig. 5), and both methods discriminated the summer propolis samples as a result of their phenolic composition. In the PCA model, the first two components comprised for 98.7 % of the total variance of the data set, suggesting that phenolic compounds seem to be an important class of metabolites for discrimination of propolis. Indeed, taking into account the improved discrimination shown in the PCA results using the UV-Vis fingerprint region of phenolic compounds, one could speculate that by targeting those compounds in propolis extracts better classification models would come about.

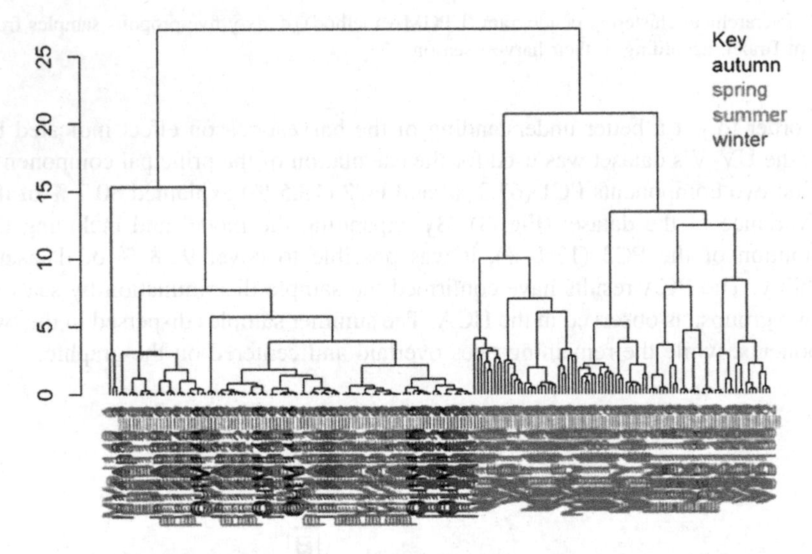


**Fig. 4** Hierarchical clustering dendogram (UPGMA method) of the fingerprint region of absorbances of phenolic compounds (UV-Vis, $\lambda$ = 280–350 ηm) of sixty five propolis samples from São Joaquim County (Santa Catarina state, southern Brazil)

These findings are of interest for the purpose of quality control processes of propolis extracts in industry, based on the fact that most of their well-known pharmacological activities rely on those secondary metabolites. In general, the majority of phytochemicals belong to the groups of phenolic compounds, alkaloids, and terpenes [12]. Nonetheless, flavonoids, phenolic acids and their ester derivatives are the major metabolites found in propolis [13]. For instance, the European propolis is characterized by their prominent amounts of flavonoids, which are not

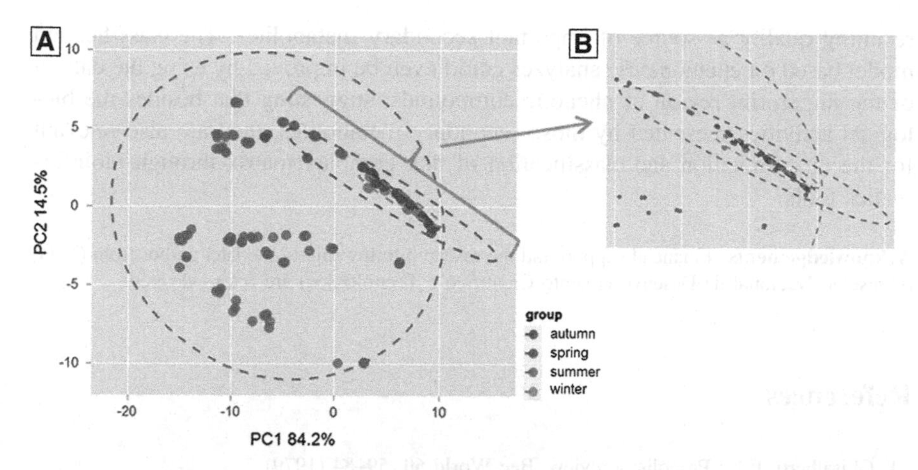


**Fig. 5** A - Principal components analysis scores scatter plot of the UV-Vis spectral profile ($\lambda = 280$–$350$ ηm) of sixty five propolis samples from southern Brazil (São Joaquim County, Santa Catarina state). B - Amplification to the overlapping samples at the PCA

often found in tropical samples [14]. In the later, prenylated phenylpropanoids are often present, the best known is the 3-(4-hydroxy-3, 5-di (3-methyl-2-butenyl) phenyl)-2(*E*)-propenoic acid [15], a high valuable compound (€ 230/10 mg) also known as Artepillin C, who has been patented for the treatment of tumors [16, 17]. *Baccharis dracunculifolia* is a native plant to Brazil commonly found in Minas Gerais state (southeastern Brazil) and source of a green resin, the main source of Artepillin C [18, 19]. Considering the interaction between *B. dracunculifolia* and *A. mellifera*, the best period to produce propolis rich in Artepillin C is from December to April, i.e., summer time in south hemisphere [20]. In this context, one can note the importance of identifying the seasonality effect on the propolis chemical profile and its resulting quality as source of important secondary metabolites. Finally, in despite the fact that UV-Vis scanning spectrophotometry is a fast, cheap and reliable analytical technique, the amount of data afforded makes unfeasible the selection of propolis samples according to their spectral profile by visual inspection of the spectra, turning the bioinformatics tools mandatory for the recovery of important features for the classification of heterogeneous samples into similar groups.

## 4 Conclusions

The spectrophotometric profile approach associated with chemometric analyses (PCA and HCA), allowed identifying a different grouping pattern in samples of propolis produced during the summer season over the other seasons, inferring the importance of the seasonality effect on the propolis chemical profile and its resulting quality as source of important secondary metabolites. The classification model based on chemometric analyzes could even be improved by using the dataset of the fingerprint region of phenolic compounds, suggesting that besides the biological activities presented by those secondary metabolites, they are also relevant for the discrimination and classification of that complex matrix through bioinformatics tools.

**Acknowledgements** Financial support and the research fellowship to the later author from CNPq (Conselho Nacional de Desenvolvimento Científico e Tecnológico) are acknowledged.

## References

1. Ghisalberti, E.L.: Propolis: a review. Bee World **60**, 59–84 (1979)
2. Greenaway, W., Scaysbrook, T., Whatley, F.R.: The composition and plant origins of propolis. Bee World **71**, 107–118 (1990)
3. Marcucci, M.C.: Biological and therapeutic properties of chemical propolis constituents. Quím. Nova **19**, 529–336 (1996)
4. Banskota, A.H., Tezuka, Y., Kadota, S.: Recent progress in pharmacological research of propolis. Phytotherapy Res. **15**, 561–571 (2001)
5. Park, Y.K., Alencar, S.M., Aguiar, C.L.: Botanical origin and chemical composition of Brazilian propolis. J. Agric. Food Chem. **50**, 2502–2506 (2002)
6. Zeggio, A. R. S.: Estudo da própolis catarinense, com ênfase na sua composição fenólica: uma estratégia à tipificação regional. Qualificação (Doutorado) – Pós Graduação em Biotecnologia e Biociências, Universidade Federal de Santa Catarina (2014)
7. Popova, M., Bankova, V., Butovska, D., Petkov, V., Nikolova-Damyanova, B., Sabatini, A. G., Marcazzan, G.L., Bogdanov, S.: Validated methods for the quantification of biologically active constituents of poplar-type propolis. Phytochem. Anal. **15**, 235–240 (2004)
8. Hanson, A.B.: ChemoSpec: an R package for chemometric analysis of spectroscopic data and chromatograms (Package Version 1.51-0) (2012)
9. Beleites, C.: Import and export of spectra files. Vignette for the R package hyperSpec (2011)
10. Bachereau, F., Marigo, G., Asta, J.: Effects of solar radiation (UV and visible) at high altitude on CAM-cycling and phenolic compound biosynthesis in *Sedum album*. Physiol. Plant. **104**, 203–210 (1998)
11. Vicini, L., Souza, A.M.: Análise multivariada da teoria à prática. Santa Maria, Brasil: UFSM 1 (2005)
12. Kennedy, D.O., Wightman, E.L.: Herbal extracts and phytochemicals: plant secondary metabolites and the enhancement of human brain function. Adv. Nutr. **2**, 32–50 (2011)
13. Banskota, A.H., Tezuka, Y., Prasain, J.K., Matsushige, K., Saiki, I., Kadota, S.: Chemical constituents of Brazilian propolis and their citotoxic activities. J. Nat. Prod. **61**, 896–900 (1998)
14. Bankova, V.: Chemical diversity of propolis and the problem of standardization. J. Ethnopharmacol. **100**, 114–117 (2005)
15. Bankova, V.S., Castro, S.L., Marcucci, M.C.: Propolis: recent advances in chemistry and plant origin. Apidologie **31**, 3–15 (2000)
16. Park, Y.K., Paredes-Guzman, J.F., Aguiar, C.L., Alencar, S.M., Fujiwara, F.Y.: Chemical constituents in Baccharis dracunculifolia as the main botanical origin of southeastern Brazilian propolis. J. Agric. Food Chem. **52**, 1100–1103 (2004)

17. Kumazawa, S., Yoneda, M., Shibata, I., Kanaeda, J., Hamasaka, T., Nakayama, T.: Direct evidence for the plant origin of Brazilian propolis by the observation of honeybee behavior and phytochemical analysis. Chem. Pharm. Bull. **51**, 740–742 (2003)
18. Hata, T., Tazawa, S., Ohta, D., Rhyu, M., Misaka, T., Ichihara, K.: Artepillin C, a major ingredient of Brazilian propolis, induces a pungent taste by activating TRPA1 Channels. PLoS ONE **7**, 1–9 (2012)
19. Alencar, S.M., Aguiar, C.L., Paredes-Guzmán, J.F., Park, Y.K.: Chemical composition of *Baccharis dracunculifolia*, the botanical source of propolis from the states of São Paulo and Minas Gerais. Brazil. Ciên. Rural **35**, 909–915 (2005)
20. Sforcin, J.M., Sousa, J.P.B., Filho, A.A., Bastos, J.K., Búfalo, M.,C., Tonuci, L.R.S.: *Baccharis dracunculifolia*: uma das principais fontes vegetais da própolis brasileira. São Paulo, Brasil: Editora Unesp, vol. 1 (2012)

# UV-Visible Spectrophotometry-Based Metabolomic Analysis of *Cedrela Fissilis* Velozzo (Meliaceae) Calluses - A Screening Tool for Culture Medium Composition and Cell Metabolic Profiles

**Fernanda Kokowicz Pilatti, Christopher Costa, Miguel Rocha, Marcelo Maraschin and Ana Maria Viana**

**Abstract** In plant cell cultures aiming at the production of secondary metabolites of industrial interest, the culture medium composition is a decisive step for obtaining cell growth and high yields of the target compound(s). A rapid and reliable methodology for screening metabolic responses to medium composition is fundamental for the development of this biotechnological field. Following this approach, UV-Vis scanning spectrophotometry of callus extracts and their spectra pre-processing, univariate and multivariate analysis were tested in the present work. The results obtained successfully discriminated the culture media investigated and shed light on what metabolic pathways might be responsible for the differences among the callus cultures' metabolic profiles.

**Keywords** Metabolic profiling • *Cedrela fissilis* • UV-Visible spectrophotometry • R language • Unsupervised methods

## 1 Introduction

*Cedrela fissilis* Vellozo (Meliaceae) is a timber tree species native to the Rainforest (South Brazil), rich in secondary metabolites, mainly terpenes, with inhibitory activity of certain enzymes of *Leishmania* spp, *Plasmodium* spp, and *Trypanosoma*

F.K. Pilatti (✉) · M. Maraschin
Plant Morphogenesis and Biochemistry Laboratory, Federal University of Santa Catarina, Florianópolis, Brazil
e-mail: fernanda.kop@gmail.com

C. Costa · M. Rocha
Centre Biological Engineering School of Engineering, University of Minho, Braga, Portugal

A.M. Viana
Botany Department, Federal University of Santa Catarina, Florianópolis, Brazil



R. Overbeek et al. (eds.), *9th International Conference on Practical Applications of Computational Biology and Bioinformatics*, Advances in Intelligent Systems and Computing 375, DOI 10.1007/978-3-319-19776-0_4

spp, insecticide properties, and folk medicine applications [1–5]. However, its medicinal potential is not exploited due to the risk of species extinction in its natural habitat and the difficulties for its *in vitro* culture or in nurseries.
Reports on shoot culture and on *in vitro* conservation of *C. fissilis* have been published and cell culture studies showed variable volatile oils composition in callus cultures according to the light conditions [6–8]. However, the lack of more information regarding the suitable culture conditions and the difficulty to establish consistent callus culture systems for *C. fissilis* have been hampering the biotechnological applications of that species for secondary metabolites production.
In the past years, metabolomics has been used in the discrimination and classification of distinct biological samples, for the identification of biomarkers in pharmaceutical, food, and beverage industries and for biological studies in microbiological, plant and animal fields [9–11]. The metabolic fingerprinting, one of the metabolomics branches, is a high-throughput, rapid, global analysis to discriminate between samples of different biological status or origin, where metabolites quantification and identification are not employed [12, 13]. Optical spectroscopic methods using infrared (IR) and UV-Visible (UV-Vis) wavelengths are rapid, cheap, and provide metabolic fingerprints that can be processed enabling pattern recognition between samples, e.g., similar or discrepant traits. UV-Vis scanning spectrophotometry requires little sample amount and preparation and rapidly provides valuable and robust information about the presence of particular classes of metabolites, such as flavonoids, carotenoids, chlorophylls and chalcones [14]. Besides sample preparation and data acquisition, a metabolomics workflow includes spectra pre-processing, for background noise filtering and data normalization. Given the complexity of the datasets, in order to identify global differences among samples the most popular approaches include unsupervised methods such as principal component analysis (PCA), hierarchical clustering (HCA), and K-means clustering [15, 16].
The present study aimed to investigate the application of UV-Vis scanning spectrophotometry and computational statistical analysis to assess the effect of carbon sources and glutamine on metabolic profiles of *C. fissilis* calluses.

## 2 Materials and Methods

### *2.1 Plant Material and* in Vitro *Cell Cultures*

The experiments were carried out with *C. fissilis* seedlings produced under sterile conditions [6]. Seeds originating from São Paulo State were harvested from selected trees by Instituto Florestal (São Paulo, Brazil) and held at 5 °C until use. The cultures were kept at 25 °C, 16 h photoperiod, and photon flux of 20–25 μmol. $m^{-2}.s^{-1}$ at plant level, supplied by Philips TDL fluorescent light tubes. Cotyledonary node cuttings (0.8–1.0 cm length) explants from 8-week seedlings were cultured on

MS medium [17] supplemented with 0.2 % (w/v) Phytagel®, 2.5 μM benzylaminopurine, 5 μM naphthalene acetic acid, 118 mM of either sucrose, glucose or fructose and glutamine (0 or 2.73 mM). There were 15 replicates per treatment. Data of fresh callus weight and dry weight were recorded after 8 weeks culture.

## 2.2 *Metabolic Profiling by UV-Vis Scanning Spectrophotometry*

Callus samples (2.0 g, fresh weight) were grounded and extracted in 3 volumes of chloroform: methanol (1:1, v/v) solution for 15 min at room temperature. Extracts were then filtered, dried at 60 °C, re-solubilized in 4 mL methanol and centrifuged (3000 rpm, 5 min). UV-Vis scanning spectrophotometry was performed for absorbance recording in the range of 200–800 ηm wavelengths using a UV-Vis spectrophotometer (Hitachi U-1800). Five scans per treatment were obtained with a 2 ηm/data point resolution.

## 2.3 *Statistical Analysis*

The UV-Vis spectra dataset was preprocessed for baseline correction, smoothing interpolation (Savitzky-Golay methods), followed by first derivative calculation. The data matrix was analyzed using scripts written in R language (v. 3.1.1) using tools defined by our research group (freely available in the following web site http://darwin.di.uminho.pt/metabolomicspackage) and some functions from the packages devtools and HyperSpec [18, 19]. These data were used for fold-change and *t*-test univariate analyses, considering glutamine-containing and glutamine-free culture medium as the two groups to compare. Principal component (PCA), hierarchical clustering (HCA) and K-means multivariate analyses were also performed. The raw data, additional chemometric analysis and the data analysis report automatically generated using the features provided by R Markdown are available in supplementary material (http://darwin.di.uminho.pt/metabolomicspackage/cedrella.html).

# 3 Results and Discussion

The raw UV-Vis spectra are usually irregular and visually polluted due to the interferences such as cuvette composition and solvent instability. The smoothing interpolation demonstrated to be important for visual melioration of the spectra as shown in Fig. 1.

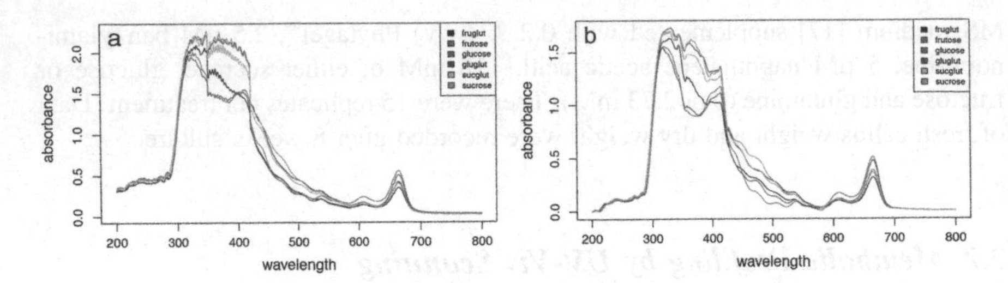


**Fig. 1** UV-Vis spectral profiles (λ = 200–800 ηm) before (a) and after (b) baseline correction by Savitzky–Golay smoothing method

Fold-change analysis detected important discriminating points nearby 210 and 580 ηm (data shown on report). HCA separated sucrose, sucrose plus glutamine and glucose plus glutamine treatments in one branch, and glucose, fructose and fructose plus glutamine ones on the other branch (data shown on report). A similar separation was observed by K-means analysis. However, a better model resulted when using the first derivative dataset as the *t*-test revealed to be the 260–270, 390, and 470 ηm wavelengths important for sample discrimination according to their metabolic profiles ($p < 0.05$, data shown on report). Importantly, flavonoid and carotenoid compounds are found to absorb at 390 and 470 ηm, respectively [14]. PCA results (first two principal components, Cumulative Proportion = 0.6725) enabled the visual identification of three clusters: (i) fructose and glucose; (ii) fructose plus glutamine and sucrose plus glutamine; (iii) sucrose and glucose plus glutamine (Fig. 2a). HCA clearly separated fructose and glucose treatments from the other ones (Fig. 2b). Sucrose and glucose plus glutamine showed to be closely related, as determined by PCA analysis, but also shared similarities with fructose plus glutamine and sucrose plus glutamine. Interestingly, K-means analysis revealed the same object clustering (data shown on report), demonstrating robustness in these analytical tools and reinforcing the importance of the first derivative transformation of the dataset as pre-processing step.

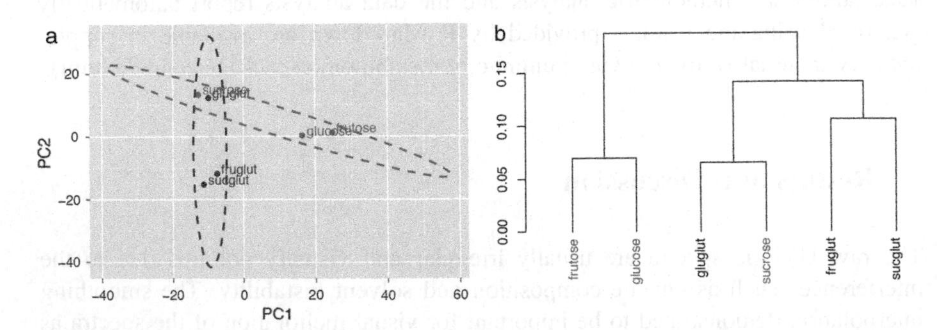


**Fig. 2** (a) PCA and (b) HCA results on Savitzky–Golay first derivative transformation of the UV-Vis spectra dataset (λ = 200–800 ηm)

By visual inspection, it can be noted that the entire UV-Vis spectral window analyzed (200–800 ηm) can be divided into two minor windows (200–450 and 630–700 ηm) according to the most relevant absorbances detected. The former spectral region encompasses the signals of absorbances of important secondary metabolites such as phenolic acids, flavonoids, and carotenoids, as the second one is usually attributed to chlorophylls [14]. In order to check the eventual effect of those classes of metabolites on the performance of the analysis models, the UV-Vis dataset was split into two subsets, i.e., 200–450 ηm (subset 1) and 630–700 ηm (subset 2) and re-analyzed.

When applied to subset 1 (200–450 ηm), *t*-test results indicated wavelengths around 300, 340 and 380 ηm as the most relevant for the differences ($p < 0.05$, data shown on report). In K-means and HCA analysis treatments were clustered in two groups: glucose plus glutamine, sucrose, and sucrose plus glutamine; fructose, fructose plus glutamine and glucose. The first derivative transformation applied to the dataset was not able to improve the discrimination of the groups in HCA and K-means, but improved PCA (data shown on report). It is possible that the complex mixture of compounds with absorbances in the addressed range of wavelengths impairs clustering. Still, the results sustain the similarity between fructose and glucose, and glucose plus glutamine and sucrose.

The smoothing method of Savitzky–Golay suppressed a minor peak (around 655 ηm) at the left side of the biggest peak in subset 2 (630–700 ηm) and apparently omitted important information for PCA results (data shown on report). *t*-test revealed wavelengths around 630 ηm as the most important in discriminating the treatments ($p < 0.05$, data shown on report). HCA and K-means analyses of this subset shown similar clustering patterns to the ones observed in the entire spectral window (data shown on report). Furthermore, these clusters match with biochemical quantification of chlorophylls, in which callus grown in culture medium with either glucose or fructose, or fructose plus glutamine had the lowest contents of this pigment.

The results highlight the importance of pre-processing the UV-Vis spectra on metabolomics studies and suggest the relevance of the comparison of different statistical methods in order to obtain more robust results. In diverse fields as clinical chemistry, drug analysis and pollution control, smoothing techniques and derivative transformation of UV-Vis spectra, followed by its multi-component analysis has proved to be useful for resolving mixtures of analytes of interest [20]. However, to our knowledge, this approach is novel in plant cell culture. Although it was possible to identify statistical differences and discriminating samples when the dataset was analyzed globally or divided into subsets, the clustering obtained in all the cases confirmed similarity between culture media containing sucrose and glucose plus glutamine, fructose and glucose, and fructose plus glutamine and sucrose plus glutamine.

According to Harborne [14], cytochromes usually are detected at the region of visible light around 545 and 605 ηm, comprehending the wavelengths at 580–590 ηm, which were relevant (fold-change analysis) to discriminate the samples. Besides, the absorbances at 470 and 630–640 ηm also contributed (*t*-test,

$p < 0.05$) for the grouping of samples and have been attributed to the chlorophylls and accessory pigments of photosynthesis (i.e., carotenoids – 470 ηm). Thus, one could speculate that the carbon and nitrogen sources added to the culture medium modulated metabolic pathways in *C. fissilis* callus cultures associated with bioenergetics cellular events. The wavelengths in region of 340–390 ηm, also pointed by *t*-test, correspond to flavonoids range of absorbance, indicating that the biosynthesis pathway for this secondary metabolite group is being activated differentially by the medium composition, which is a result of biotechnological interest.

It is well established in literature that carbon and nitrogen sources, as well as their ratios, are determinant for cell growth and differentiation, and for secondary metabolites production. The results herein presented indicate that glucose and fructose supplemented separately activated similar primary and secondary metabolic routes. In its turn, sucrose seems to affect differently the metabolic pathways. This can be explained by the fact that sucrose hydrolysis releases to the cell equal amounts of glucose and fructose, resulting in a cellular metabolic status where those monosaccharides differently interact with other metabolites and regulate cellular processes. Besides considering the influence of differential uptake of glucose and fructose from sucrose hydrolysis in cell metabolism, one could expect metabolic profiles distinct in any extension, taking into account the regulatory and signaling roles of those simple sugars. Indeed, since glucose and fructose are not solely energy sources for the cells, also playing important roles in cell signaling and plant growth and development, it is reasonable to assume that the carbon sources directly determine metabolic response [21, 22].

The other cluster observed in our study was composed by the treatments with glutamine supplementation and sucrose solely. Nitrogen is usually supplied in culture medium as nitrate or ammonium, the first one being converted to the latest, and this one being used for the synthesis of glutamine, which is a precursor of glutamate, a key molecule for other amino acids synthesis, nucleotides and chlorophylls [23, 24]. The MS culture medium used in the present study contains inorganic nitrogen in the form of ammonium nitrate. The addition of glutamine greatly impacted the cell metabolism, being plausible to hypothesize that amide acted as carbon source for energetic purposes similarly to the other sugars supplemented to the medium. Indeed, it has been demonstrated that glutamine supplementation in culture medium can be preferred as carbon source to plant cells over glucose in TCA cycle, affecting the imbalance of all the pathways linked to it, and, as a consequence, the cellular carbohydrates status. Furthermore, alanine resultant from the glutamine catabolism may be directed to gluconeogenesis, feeding back and overloading TCA cycle. Moreover, the presence of glutamine and other amino acids in culture medium inhibits nitrite and ammonium uptake, and may slow the glycolytic pathway, which also changes energy status and cell signaling, therefore providing another possible explanation for the discrimination of the callus cultures metabolic profiles herein presented [25, 26].

# 4 Conclusions

UV-Vis scanning spectrophotometry and computational methods used in this work proved to be suitable for the proposed goal. The pre-processing of the UV-Vis dataset proved to be an important step, affecting subsequent analysis. Univariate and multivariate analysis enabled satisfactory discrimination of the metabolic profiles investigated, as well as indications for further metabolic target investigation on how carbon sources and glutamine supplementation affect *C. fissilis* callus cultures. The information obtained, useful for cell culture medium optimization aiming at secondary metabolites production relies mostly on the bioinformatics tools applied to the complex UV-Vis dataset, both in its original form and in its pre-processed one.

**Acknowledgements:** To CAPES (Coordination for the Improvement of Higher Education Personnel) for financial support, Post-Graduation Program in Cell and Developmental Biology (Federal University of Santa Catarina) and BIOSYSTEMS research group (University of Minho).

# References

1. Goulart, H.R., Kimura, E.A., Peres, V.J., Couto, A.S., Aquino Duarte, F.A., Katzin, A.M.: Terpenes arrest parasite development and inhibit biosynthesis of isoprenoids in *Plasmodium falciparum*. Antimicrob. Agents Chemother. **48**, 2502–2509 (2004)
2. Leite, A.C., Bueno, F.C., Oliveira, C.G., Fernandes, J.B., Da Vieira, P.C., Silva, M.F.G.F., Bueno, O.C., Pagnocca, F.C., Hebling, M.J.A., Bacci Jr, M.: Limonoids from *Cipadessa fruticosa* and *Cedrela fissilis* and their insecticidal activity. J. Braz. Chem. Soc. **16**, 1391–1395 (2005)
3. Ambrozin, A.R.P., Leite, A.C., Bueno, F.C.: Limonoids from andiroba oil and *Cedrela fissilis* and their insecticidal activity. J. Braz. Chem. Soc. **3**, 542–547 (2006)
4. Ramos, D.F., Leitão, G.G., Costa, F.N., Abreu, L., Villarreal, J.V., Leitão, S.G., Fernández, S. L.S., Da Silva, P.E.A.: Investigation of the antimycobacterial activity of 36 plant extracts from the Brazilian Atlantic Forest. Braz. J. Pharm. Sci. **44**, 669–674 (2008)
5. Leite, A.C., Ambrozin, A.R.P., Castilho, M.S., Vieira, P.C., Fernandes, J.B., Oliva, G., Silva, M.F.G.F., Thiemann, O.H., Lima, M.I., Pirani, J.R.: Screening of *Trypanosoma cruzi* glycosomal glyceraldehyde-3-phosphate dehydrogenase enzyme inhibitors. Rev. Bras. Farmacogn. **19**, 1–6 (2009)
6. Nunes, E.C., Castilho, C.V., Moreno, F.N., Viana, A.M.: In vitro culture of *Cedrela fissilis* Vellozo (Meliaceae). Plant Cell Tiss. Org. **70**, 259–268 (2002)
7. Nunes, E.C., Benson, E.E., Oltramari, A.C., Araujo, P.S., Moser, J.R., Viana, A.M.: In vitro conservation of *Cedrela fissilis* Vellozo (Meliaceae), a native tree of the Brazilian Atlantic Forest. Biodivers. Conserv. **12**, 837–848 (2003)
8. Nunes, E.C., Laudano, W.L.S., Moreno, F.N., Castilho, C.V., Mioto, P., Sampaio, F.L., Bortoluzi, J.H., Benson, E.E., Pizolatti, M.G., Carasek, E., Viana, A.M.: Micropropagation of *Cedrela fissilis* Vell. (Meliaceae). In: Jain, S.M., Häggman, H. (ed.) – Protocols for Micropropagation of Woody Trees and Fruits, pp. 221–235. Springer, The Netherlands (2007)
9. Nielsen, J., Jewett, M.C.: Metabolomics - A powerful pool in systems biology. Springer-Verlag, Berlin, Heidelberg (2007)

10. Villas-Bôas, S.G., Roessner, U., Hansen, M.A.E., Smedsgaard, J., Nielsen, J.: Metabolome analysis: An introduction. Wiley, New Jersey (2007)
11. Putri, S.P., Nakayama, Y., Matsuda, F., Uchikata, T., Kobayashi, S., Matsubara, A., Fukusaki, E.: Current metabolomics: practical applications. J. Biosci. Bioeng. **115**, 579–589 (2013)
12. Fiehn, O.: Combining genomics, metabolome analysis, and biochemical modelling to understand metabolic networks. Comp. Funct. Genomics **2**, 155–168 (2001)
13. Dunn, W.B., Ellis, D.I.: Metabolomics: current analytical platforms and methodologies. Trend. Anal. Chem. **24**, 285–294 (2005)
14. Harborne, J.B.: Phytochemical Methods, 3rd edn. Chapman & Hall, London (1998)
15. Sumner, L.W., Mendes, P., Dixon, R.A.: Plant metabolomics: large-scale phytochemistry in the functional genomics era. Phytochemistry **62**, 817–836 (2003)
16. Xia, J., Mandal, R., Sinelnikov, I.V., Broadhurst, D., Wishart, D.S.: MetaboAnalyst 2.0 - a comprehensive server for metabolomic data analysis. Nucl. Acids Res. **37**, W652–W660 (2009)
17. Murashige, T., Skoog, F.: A revised medium for rapid growth and bioassays with tobacco tissue cultures. Physiol. Plant. **15**, 473–497 (1962)
18. Wickham, H., Chang, W., RStudio, R Core team: Tools to Make Developing R Packages Easier (2015)
19. Beleites, C.: Import and Export of Spectra Files. Vignette for the R package hyperSpec (2011)
20. Bosch Ojeda, C., Sanchez Rojas, F.: Recent developments in derivative ultraviolet/visible absorption spectrophotometry. Anal. Chim. Acta **518**, 1–24 (2004)
21. Gamborg, O.L., Murashige, T., Thorpe, T.A., Vasil, I.K.: Plant tissue culture media. In vitro. **12**, 473–478 (1976)
22. Rolland, F., Baena-Gonzalez, E., Sheen, J.: Sugar sensing and signaling in plants: conserved and novel mechanisms. Annu. Rev. Plant Biol. **57**, 675–709 (2006)
23. Miflin, B.J., Habash, D.Z.: The role of glutamine synthetase and glutamate dehydrogenase in nitrogen assimilation and possibilities for improvement in the nitrogen utilization of crops. J. Exp. Bot. **53**, 979–987 (2002)
24. Forde, B.G., Lea, P.J.: Glutamate in plants: metabolism, regulation, and signalling. J. Exp. Bot. **58**, 2339–2358 (2007)
25. Flores-Samaniego, B., Olivera, H., Gonzalez, A.: Glutamine synthesis is a regulatory signal controlling glucose catabolism in *Saccharomyces cerevisiae*. J. Bacteriol. **175**, 7705–7706 (1993)
26. DeBerardinis, R.J., Mancuso, A., Daikhin, E., Nissim, I., Yudkoff, M., Wehrli, S., Thompson, C.B.: Beyond aerobic glycolysis: transformed cells can engage in glutamine metabolism that exceeds the requirement for protein and nucleotide synthesis. PNAS **104**, 19345–19350 (2007)

# An Integrated Computational Platform for Metabolomics Data Analysis

**Christopher Costa, Marcelo Maraschin and Miguel Rocha**

**Abstract** The field of metabolomics, one of the omics technologies that have recently revolutionized biological research, provides multiple challenges for data analysis, that have been addressed by several computational tools. However, none addresses the multiplicity of existing techniques and data analysis tasks. Here, we propose a novel R package that provides a set of functions for metabolomics data analysis, including data loading in different formats, pre-processing, univariate and multivariate data analysis, machine learning and feature selection. The package supports the analysis of data from the main experimental techniques, integrating a large set of functions from several R packages in a powerful, yet simple to use environment, promoting the rapid development and sharing of data analysis pipelines.

**Keywords** Metabolomics R NMR MS IR UV-vis package data analysis

## 1 Introduction

Metabolomics can be defined as the identification and quantification of all intracellular and extracellular metabolites with low molecular mass. It offers valuable tools in functional genomics and, more globally, in the characterization of biological systems [10]. Applications include studying metabolic systems, measuring biochemical phenotypes, understanding and reconstructing networks, discriminating between samples, identifying biomarkers of disease, analyzing food and beverages, studying plant physiology or fostering drug discovery [9, 10, 15].

Unlike transcriptomics and proteomics which are based on polymers, metabolites have a large variance in chemical structures and properties, making difficult the development of high-throughput techniques and reducing the number of molecules

C. Costa (✉) · M. Rocha
CEB - Centre Biological Engineering, University of Minho, Braga, Portugal

M. Maraschin
Plant Morphogenesis and Biochemistry Laboratory, Federal University of Santa Catarina, Florianopolis, Brazil



R. Overbeek et al. (eds.), *9th International Conference on Practical Applications of Computational Biology and Bioinformatics*, Advances in Intelligent Systems and Computing 375, DOI 10.1007/978-3-319-19776-0_5

that can adequately be measured in a sample [15]. Thus, there are several experimental techniques to obtain metabolomics data. In this work, the main approaches will be supported, namely: Nuclear Magnetic Resonance (NMR), Liquid Chromatography (LC) and Gas Chromatography (GC) with Mass Spectrometry (MS), Infrared (IR) and Ultraviolet-visible (UV-vis) spectroscopies. GC/LC-MS and NMR are more robust techniques and they are frequently employed in the analysis and quantification of the metabolome. IR and UV-vis techniques have the advantage of being simpler and less expensive, providing a complementary view of that provided by GC/LC-MS or NMR.

Two distinct approaches can be chosen for planning and executing a metabolomics experiment. The former, a chemometrics approach or metabolic fingerprinting, makes direct use of spectra or peaks lists, and the analysis typically addresses sample discrimination. The latter, known as metabolic target analysis or profiling, focuses on the identification and quantification of compounds in the sample, using that information to run the analysis. The general workflow of an experiment generally consists in the steps of sample preparation, data acquisition, preprocessing, data analysis and interpretation (see Fig. 1). Once the samples are prepared and data is acquired, it will be preprocessed to correct some issues and improve the performance of the next step, data analysis.

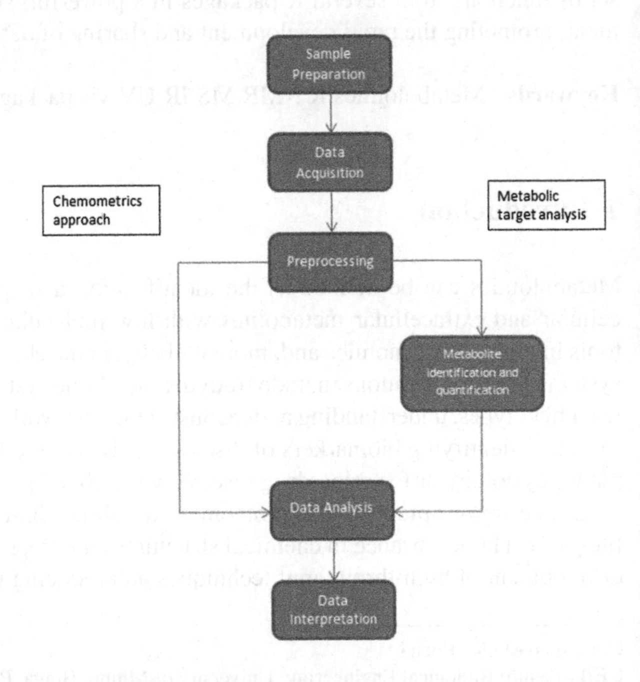


**Fig. 1** General workflow of a metabolomics experiment

Metabolomics brings important challenges regarding the extraction of relevant knowledge from large amounts of data [14]. Indeed, omics data have promoted the development and adaptation of numerous methods for data analysis. Some of the main available tools are summarized in Table 1. The most comprehensive tool found was the web-based suite MetaboAnalyst [16], which provides a number of tools with a user-friendly interface. It is, however, limited to the provided environment and available configuration parameters, being difficult to extend and define personalized workflows and does not cover spectral data. On the other hand, a number of R packages have been developed, from which we emphasize *xcms* (from Bioconductor), *hyperSpec*, *ChemoSpec* [2, 7], which provide valuable functions for all types of metabolomics data analysis. However, these packages are not well integrated with each other, and therefore it is not straightforward to combine their functionalities.

**Table 1** Available free tools for metabolomics data

| Name | URL | Short description |
|---|---|---|
| MetaboAnalyst | http://www.metaboanalyst.ca | Web application to analyze metabolomic data |
| *HyperSpec* | http://hyperspec.r-forge.r-project.org | R package to handle spectral data and metadata |
| ChemoSpec | http://cran.r-project.org/web/packages/ChemoSpec | R package to handle spectral data |
| Metabolomic package | http://cran.open-source-solution.org/web/packages/Metabonomic/ | R-package + GUI for metabonomic profiles |
| Speaq | https://code.google.com/p/speaq/ | Integrated workflow for robust alignment and quantitative analysis of NMR |
| Automics | https://code.google.com/p/automics/ | NMR-based spectral processing and analysis |
| MeltDB | https://meltdb.cebitec.uni-bielefeld.de | Web-based system for data analysis |
| Metabolomics | http://cran.r-project.org/web/packages/metabolomics | R functions for metabolomics statistical analysis |
| MetaP-Server | http://metabolomics.helmholtz-muenchen.de/metap2/ | Web application to analyze metabolomic data |
| Bioconductor | http://bioconductor.org/packages/release/BiocViews.html # Metabolomics | Bioconductor R packages for metabolomics (include xcms) |

In this scenario, this work aims to develop an integrated script-based software for the analysis of metabolomics data. This package is written in R, taking advantage of the available functions provided by specific metabolomics oriented packages as highlighted above, but also more general-purpose data analysis ones. The developed package will address at this stage data from MS, NMR, IR and UV-vis metabolomics experiments. The software was developed to address a wide variety of common tasks on metabolomics data analysis, providing a general workflow that can be adapted for specific case studies.

## 2 Development of the Package and Provided Functions

To achieve the objectives of this work, a package with features covering the main steps of the metabolomics data analysis workflow was developed, containing functions for the data reading and dataset creation, preprocessing and data analysis. Figure 2 shows the main modules developed.

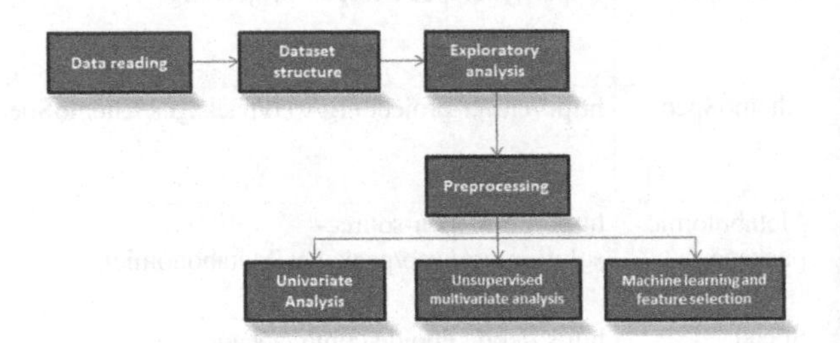


**Fig. 2** Modules developed in the package

The package contains functions that are easy to use, i.e. with few mandatory parameters, but also very flexible, since most functions have default parameters, but they also have a large number of parameters that can be used to change the default behavior. The package integrates many functions imported and sometimes adapted/ extended from other packages, integrating various packages over a unique interface. The package's functions were meant to provide abundant graphical visualization options of the results. The idea is to minimize the complexity of the code needed to make an analysis pipeline over a dataset, but also to easily allow creating variants for this analysis.

The package was developed using the R[1] environment, a free software environment for data manipulation, scientific and statistical computing and graphical visualization. RStudio[2] was the environment chosen to develop the scripts and assemble the package. Reports were made using a plugin named RMarkdown,[3] which can create easily dynamic HTML reports from annotated R code.

[1] http://www.r-project.org.

[2] http://www.rstudio.com.

[3] http://rmarkdown.rstudio.com.

The package can be accessed and installed through the ZIP file available in the package's homepage: http://darwin.di.uminho.pt/metabolomicspackage. The package comes with the full functions documentation. In the future, it will be made available on CRAN, the R community package repository.

## *2.1 Data Reading and Dataset Structure*

The types of metabolomics data supported are NMR, UV-vis, IR and MS data. A number of different file formats are supported, including Comma (or Tab) Separated Values (CSV or TSV) files, (J)DX spectra files, NetCDF, mzDATA and mzXML MS data. Metadata can be given as a CSV/TSV file. The core reading functions for (J)DX are provided by *ChemoSpec* [7]. For LC/GC-MS spectra data (NetCDF, mzDATA and mzXML), the package *xcms*[17] was used. It is also possible to load data as a peaks list, using peak alignment functions to reach a dataset in a standard tabular form. Figure 3 represents a scheme of the data reading processes.

The basic structure of a dataset is general-purpose, independent of the type of data and source. A dataset is an R list consisting of the following fields: description of the dataset, type of data, the data matrix, the metadata data frame and the labels for the x and y axis. In Fig. 4, a graphical representation of the dataset structure is provided.

## *2.2 Exploratory Analysis and Data Pre-processing*

The package includes a number of functions that allow to calculate global statistics and others to provide graphical visualization of the data. A basic visualization function allows to see the distribution of values for (a subset of) the variables in the dataset in the form of boxplots. There are also functions to plot spectral data (where variables are represented by numerical values). For visualization functions, the base graphics system of R was used.

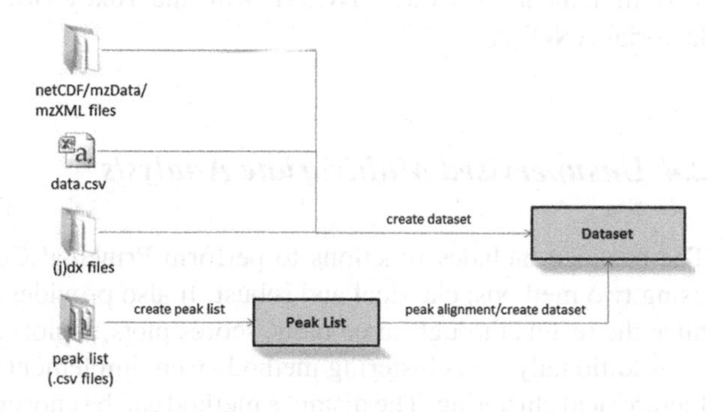


**Fig. 3** Scheme of the data reading

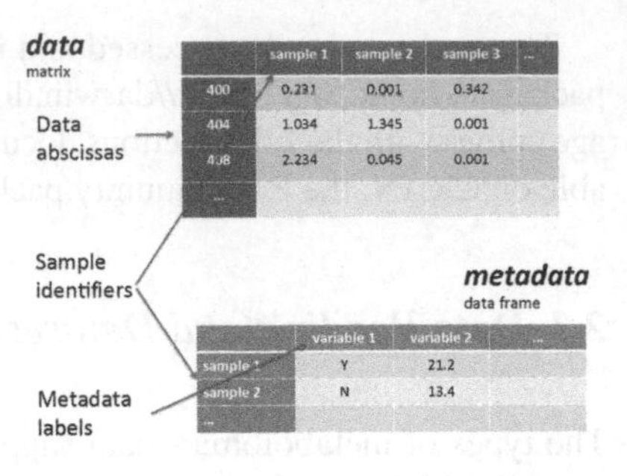


**Fig. 4** Representation of the structure of the data in a dataset

Various preprocessing methods were implemented. To extract relevant parts of a dataset, a number of functions were developed that allow to extract (or remove) relevant subsets of samples, data and metadata variables. To perform spectral corrections, methods are available for shifting correction, smoothing interpolation, multiplicative scatter correction and baseline, offset and background corrections. To treat missing values, various imputation methods were implemented. Also, methods for data normalization, transformation and scaling are provided. Finally, to remove variables with low variance, flat pattern filters are provided with distinct metrics and parameters.

## 2.3 Univariate Analysis

A number of distinct univariate analysis methods, all based in the stats R package, were implemented covering the following types types of analysis: correlations, fold changes, *t*-tests and analysis of variance (ANOVA). Indeed, there are functions to calculate the correlations between variables or samples, and the resulting matrix can be visualized as a heatmap. To perform fold change analysis, there are functions to calculate fold changes of values considering two groups of samples (as defined by a metadata variable), being possible to visualize the results in tabular and graphical forms. Also, *t*-tests can be performed and visualized in a similar way. The package also implements one-way ANOVA, with the Tukey HSD post-hoc test, and multifactorial ANOVA.

## 2.4 Unsupervised Multivariate Analysis

The package includes functions to perform Principal Component Analysis (PCA) using two methods: classical and robust. It also provides a number of ways to visualize the results through scree plots, scores plots, biplots and pairs plots.

Additionally, two clustering methods were implemented: k-means clustering and hierarchical clustering. The distance method can be chosen according to the available option from the *dist* function, as well the method used in the case of the hierarchical clustering. Clustering results can also be plotted. There are functions that create a dendrogram of the hierarchical clustering results and functions that allow to visualize the results from k-means.

### *2.5 Machine Learning and Feature Selection*

The package provides a number of functions to train, use and evaluate machine learning methods, being mostly based in the R package *caret*, covering both classification and regression methods. Also, there are functions to evaluate the importance of each variable in the models. A list of possible models and tunable parameters can be seen in: http://topepo.github.io/caret/modelList.html. The error metrics available include accuracy, Area Under the ROC Curve (AUC) and Kappa statistic for classification, and Root Mean Square Error (RMSE) and the coefficient of determination ($R^2$) for regression.

The results from model training/ optimization include the performance of the best model, the variables importance, the results of all combinations of parameters, the confusion matrices (for classification) and the final models. These can be used later for predictions or visualization, as it happens in the Partial Least Squares (PLS) case. Also, there are a number of functions to perform feature selection, i.e. determine which attributes are more valuable when applying different machine learning methods. Recursive Feature Elimination (RFE) and filtering are made available from the *caret* package.

## 3 Case Studies

Three case studies using real data will be presented to test the package and provide meaningful data analysis pipelines. The first is the analysis of metabolites concentrations of urine samples from control and cachexic cancer patients, the next is the discrimination of propolis samples from southern Brazil (using NMR and UV-vis data), and the last one is the analysis of the effect of Postharvest Physiological Deterioration (PPD) of cassava samples (IR data).

For one of the case studies, some selected results will be shown as a demonstration of some of the package's capabilities. Full data analysis reports for all cases, following distinct analysis pipelines are further provided in the following URL: http://darwin.di.uminho.pt/metabolomicspackage

### *3.1 Cachexia*

Cachexia is a complex metabolic syndrome associated with an underlying illness (such as cancer) and characterized by loss of muscle with or without loss of fat mass [5]. Improved approaches for detecting the onset and evolution of muscle wasting would help to manage wasting syndromes and facilitate early intervention [4]. As metabolites produced from tissue breakdown are likely to be a sensitive indicator of muscle wasting, urine samples were collected since several end products of muscle catabolism are specifically excreted in urine [4]. A total of 77 urine samples were collected being 47 of them patients with cachexia, and 30 control patients. All one-dimensional NMR spectra of urine samples were acquired and then the metabolites were detected and quantified, i.e. for each metabolite its concentration was measured.

### *3.2 Propolis*

Propolis is a substance produced from the collected exudates of plants (resin) by bees. The resin is masticated, salivary enzymes are added, and the partially digested material is mixed with beewax and used in the hive to seal the walls, strengthen the borders of combs, and embalm dead invaders. Recently, this product has been the subject of studies highlighting its pharmacological properties, such as the antimicrobial [1, 3], anti-oxidative [8], anti-viral [6], anti-tumoral [11, 12] or anti-inflammatory [1, 3].

It has long been known that propolis chemical composition might be strongly influenced by environmental factors peculiar to the sites of collection of a given geographic region of production, as well as by seasoning. The aim of this case study is to gain insights of important features associated to chemical composition, harvest season, and geographic origin of propolis produced in the Santa Catarina state, southern Brazil. The propolis samples used in this study for NMR data analysis were collected in the autumn (AU), winter (WI), spring (SP), and summer (SM) of 2010 from hives located in Santa Catarina state. A total of 59 samples were collected, with the distribution of samples by seasons being: SM - 16 samples, AU and SP - 15 samples, WI - 13 samples. Also, three agroecological regions were defined for the different apiaries, distributed as follows: Highlands - 12 samples, Plain - 11 samples, Plateau - 36 samples.

After preprocessing, data analysis was conducted and some of results are presented below. The one-way ANOVA results in Table 2 indicate that compounds with

**Table 2** ANOVA results for discriminating harvest seasons, with the ppms, p-values, the logarithm of the p-values, the false discovery rate and the Tukey's HSD results

| ppm | p-values | logs | fdr | Tukey |
|---|---|---|---|---|
| 4.66 | 9.585e-26 | 25.018 | 2.319e-23 | sm-au; sp-sm; wi-sm |
| 4.58 | 3.385e-17 | 16.470 | 4.096e-15 | sm-au; sp-sm; wi-sm |
| 4.55 | 6.092e-14 | 13.215 | 4.915e-12 | sm-au; sp-au; wi-au; sp-sm; wi-sm |
| 4.63 | 1.044e-13 | 12.981 | 6.316e-12 | sm-au; sp-sm; wi-sm |
| 4.71 | 2.083e-13 | 12.681 | 1.008e-11 | sm-au; sp-sm; wi-sm |

anomeric structural moieties appear to have a significant effect on the discrimination of propolis samples over the seasons, because all the main resonances selected occur at the anomeric spectral region (3.00–5.50 ppm).

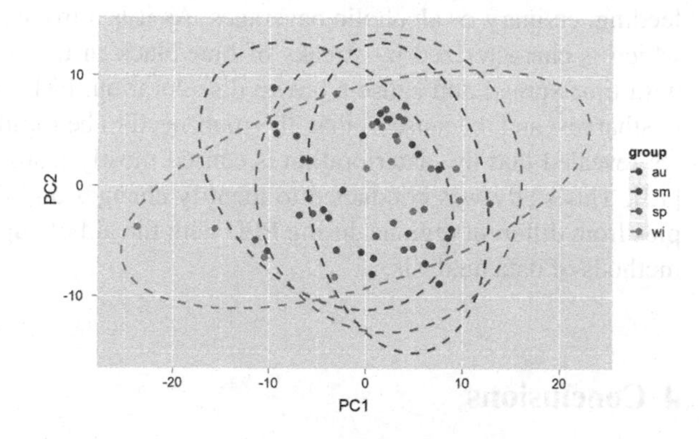


**Fig. 5** PCA scores plot (PC1 and PC2) grouped by the harvest seasons

PCA results revealed a low percentage of variance explained by the first principal components. Figure 5 shows that there is significant overlapping on the first two components. On the other hand, results from machine learning models trained to discriminate the harvest seasons are shown in Table 3 with 10-fold cross-validation with 10 repetitions. The accuracy, Kappa statistic and AUC results were quite good, which can reveal the features that are significant to discriminate the harvest seasons and to successfully predict new samples. A 3D plot of the first 3 components of the PLS model is shown in Fig. 6, showing a clear separation of the classes.

**Table 3** Classification models result for discriminating the harvest seasons

| Model | Accuracy | Kappa statistic | AUC |
|---|---|---|---|
| PLS | 86.78 % | 82.10 % | 96.94 % |
| Random forest | 84.56 % | 79.08 % | 94.51 % |

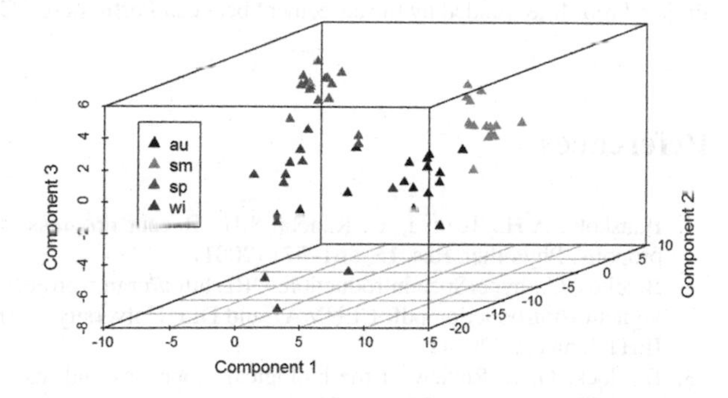


**Fig. 6** 3D plot of the first 3 components of the PLS model with the harvest seasons

### 3.3 *Cassava*

Cassava is widely cultivated in tropical and subtropical regions for its starchy tuberous root, a great source of carbohydrates. It has a variety of applications, like animal feeding, culinary or alcoholic beverages. As it is a tropical root, it undergoes PPD, which is characterized by streaks of blue/black in the root vascular tissue, which with time spread and cause a brown discoloration. PPD begins quickly within 24h postharvest and, because of that, the roots need to be rapidly consumed. Some studies revealed that the deterioration is caused mostly from wound-healing responses [13]. This study was conducted to identify changes and discriminate cassava samples from different regions during PPD with the aid of supervised and unsupervised methods of data analysis.

## 4 Conclusions

To face the numerous challenges brought by metabolomics data, an R package was developed, which includes features and methods for a variety of important aspects, starting with data loading into a defined structure, various methods of preprocessing and visual exploration of the data, as well as methods for data analysis and machine learning. All these methods have been implemented and demonstrated in distinct real-world case studies. With the possibility of quickly creating and visualizing the results of analysis pipelines, the package can be used by anyone with or without a computational background. The package is quite flexible, with functions that are easy to use having default configurations, but with the possibility of configuring more specific details as users get acquainted with the potential of the functions. Therefore, the package put forward in this work will be a valuable tool for researchers in the growing field of metabolomics.

**Acknowledgments** The work is partially funded by Project 23060, PEM - Technological Support Platform for Metabolic Engineering, co- funded by FEDER through Portuguese QREN under the scope of the Technological Research and Development Incentive system, North Operational and by Project PropMine, funded by the agreement between Portuguese FCT and Brazilian CNPq.

## References

1. Banskota, A.H., Tezuka, Y., Kadota, S.H.: Recent progress in pharmacological research of propolis. Phytother. Res. **15**, 561–571 (2001)
2. Beleites, C.: hyperSpec introduction. URL http://cran.r-project.org/web/packages/hyperSpec/vignettes/introduction.pdf. CENMAT and DI3, University of Trieste Spectroscopy - Imaging, IPHT Jena e.V (2014)
3. Burdock, G.A.: Review of the biological properties and toxicity of bee propolis (propolis). Food Chem. Toxicol. **36**, 347–363 (1998)

4. Eisner, R., Stretch, C., Eastman, T., Xia, J., Hau, D., Damaraju, S., Greiner, R., Wishart, D.S., Baracos, V.E.: Learning to predict cancer-associated skeletal muscle wasting from 1h-nmr profiles of urinary metabolites. Metabolomics **7**, 25–34 (2010)
5. Evans, W.J., Morleya, J.E., Argilésa, J., Balesa, C., Baracosa, V., Guttridgea, D., Jatoia, A., Kalantar-Zadeha, K., Lochsa, H., Mantovania, G., Marksa, D., Mitcha, W.E., Muscaritolia, M., Najanda, A., Ponikowskia, P., Fanellia, F.R., Schambelana, M., Scholsa, A., Schustera, M., Thomas, D., Wolfea, R., Anker, S.D.: Cachexia: a new definition. Clin. Nutr. **27**, 793–799 (2008)
6. Gekker, G., Hu, S., Spivak, M., Lokensgard, J.R., Peterson, P.K.: Anti-hiv-1 activity of propolis in cd4+ lymphocyte and microglial cell cultures. J. Ethnopharmacol. **102**, 158–163 (2005)
7. Hanson, B.A.: Chemospec: an r package for chemometric analysis of spectroscopic data and chromatograms (2013)
8. Kumazawa, S., Ueda, R., Hamasaka, T., Fukumoto, S., Fujimoto, T., Nakayama, T.: Antioxidant prenylated flavonoids from propolis collected in Okinawa, Japan. J. Agric. Food Chem. **55**, 7722–7725 (2007)
9. Mozzi, F., Ortiz, M.E., Bleckwedel, J., Vuyst, L.D., Pescuma, M.: Metabolomics as a tool for the comprehensive understanding of fermented and functional foods with lactic acid bacteria. Food Res. Int. (2012)
10. Nielsen, J., Jewett, M.C.: Metabolomics: A Powerful Tool in Systems Biology. Springer, Berlin (2007)
11. Sforcin, J.M.: Propolis and the immune system: a review. J. Ethnopharmacol. **113**, 1–14 (2007)
12. Tan-No, K., Nakajima, K.T., Shoii, T., Nakagawasai, O., Niijima, F., Ishikawa, M., Endo, Y., Sato, T., Satoh, S., Tadano, K.: Anti-inflammatory effect of própolis through nitric oxide production on carrageenin-induced mouse paw edema. Biol. Pharm. Bull **29**, 96–99 (2006)
13. Uarrota, V.G., Moresco, R., Coelho, B., da Costa, E.: In: Nunes, L.A., Martins Peruch, E., de Oliveira Neubert, M., Rocha, M.M. (eds.) Metabolomics combined with chemometric tools (pca, hca, pls-da and svm) for screening cassava (manihot esculenta crantz) roots during postharvest physiological deterioration. Food Chem. **161**, 67–78 (2014)
14. Varmuza, K., Filzmoser, P.: Introduction to multivariate statistical analysis in chemometrics. (2008, CRC Press)
15. Villas-Boas, S., Roessner, U., Hansen, M.A.E., Smedsgaard, J., Nielsen, J.: Metabolome analysis: an introduction. Wiley, Hoboken (2007)
16. Xia, J., Mandal, R., Sinelnikov, I.V., Broadhurst, D., Wishart, D.S.: Metaboanalyst 2.0-a comprehensive server for metabolomic data analysis. Nucleic Acids Res. **40**, 127–33 (2012)
17. Smith, C.A., Want, E.J., O'Maille, G., Abagyan, R., Siuzdak, G.: XCMS: processing mass spectrometry data for metabolite profiling using nonlinear peak alignment, matching and identification. Anal. Chem. **78**, 779–787 (2006)

# Compound Identification in Comprehensive Gas Chromatography—Mass Spectrometry-Based Metabolomics by Blind Source Separation

**Xavier Domingo-Almenara, Alexandre Perera, Noelia Ramírez and Jesus Brezmes**

**Abstract** Comprehensive gas chromatography - mass spectromety (GCxGC-MS) has become a promising tool in metabolomics. However, algorithms for GCxGC-MS data processing are needed in order to automatically process the data and extract the most pure information about the compounds appearing in the complex biological samples. This study shows the capability of orthogonal signal deconvolution (OSD), a novel algorithm based on blind source separation, to extract the spectra of the compounds appearing in GCxGC-MS samples. Results include a comparison between OSD and multivariate curve resolution - alternating least squares (MCR-ALS) with the extraction of metabolites spectra in a human serum sample analyzed through GCxGC-MS. This study concludes that OSD is a promising alternative for GCxGC-MS data processing.

**Keywords** Comprehensive gas chromatography · Orthogonal signal deconvolution · Multivariate curve resolution · Compound deconvolution · Independent component analysis

X. Domingo-Almenara (✉) · N. Ramírez · J. Brezmes
Metabolomics Platform, IISPV, Universitat Rovira i Virgili, Campus Sescelades, Carretera de Valls, s/n, 43007 Tarragona, Catalonia, Spain
e-mail: xavier.domingo@urv.cat

A. Perera
B2SLAB. Department d'Enginyeria de Sistemes, Automàtica i Informàtica Industrial, CIBER-BBN, Universitat Politècnica de Catalunya, Barcelona, Catalonia, Spain

X. Domingo-Almenara · N. Ramírez · J. Brezmes
CIBERDEM, Biomedical Research Networking Center in Diabetes and Associated Metabolic Disorders, Bonanova 69, 08017 Barcelona, Catalonia, Spain



R. Overbeek et al. (eds.), *9th International Conference on Practical Applications of Computational Biology and Bioinformatics*, Advances in Intelligent Systems and Computing 375, DOI 10.1007/978-3-319-19776-0_6

# 1 Introduction

One of the most important objectives of metabolomics is to detect early biochemical changes in organisms before a disease appears and provide an opportunity to find predictive biomarkers [1]. This can be achieved by performing non-targeted analysis of metabolites, since those biochemical changes are not known a priori. One of the analytical platforms used in metabolomics to detect very low molecular weight compounds is gas chromatography. Gas chromatography is used in metabolomics to separate a mixture of compounds contained in a biological sample, e.g., plasma, serum, urine, saliva, etc., with the purpose of identifying existing compounds and determine their concentration. This can be achieved by passing the mixture trough a chromatographic column. In some cases, certain compounds do not appear chromatographically separated and this situation is known as coelution. Comprehensive gas chromatography GCxGC-MS [2] was devised to minimize the coelution problem, where the compounds pass through two chromatographic columns with orthogonal polarity properties, submitting the sample to two analytical separations and leading to an increased compound separation capability. In addition, a mass spectrometer is attached, which turns GCxGC-MS in one of the best analytical tools for the analysis of volatile and semi-volatile analytes [3].
This reduction of chomatographic coelution in GCxGC-MS facilitates the computational interpretation of the chromatographic signals making the automation of this task more feasible. However, compounds in the sample appear at low concentrations and different sources of noise derived from the instrument and the sample biological matrix may interfere with the correct identification of the compounds. In the same way, GCxGC-MS generates large quantity of data and its interpretation can not be conducted manually. In that sense, GCxGC-MS data processing algorithms are needed to turn the chromatographic signals into interpretable biological information.
As reviewed in [4], some of the existing data processing algorithms that can be applied to resolve mixtures in comprehensive gas chromatography include PARAFAC [5] and multivariate curve resolution - alternating least squares (MCR-ALS) [6]. Contrarily to MCR, PARAFAC can be only applicable to a three-way data set, i.e., PARAFAC can not resolve a single GCxGC-MS sample.
Independent component analysis (ICA) [7] is a blind source separation (BSS) technique capable of separating linearly mixed sources. ICA is widely applied to the resolution of spectroscopy mixtures and some studies report the use of ICA to resolve mixtures in chromatographic data [8–10]. More recently, a combination of ICA and PCA known as orthogonal signal deconvolution (OSD) [11] has been applied to resolve mixtures in GC-MS. Any of these aforementioned BSS techniques have been tested yet for GCxGC-MS data processing.
In this paper we show the capability of OSD to resolve and identify the compounds appearing in a GCxGC-MS chromatogram. The performance of OSD was tested by comparing the spectra extraction of 16 metabolites appearing in a human serum sample with the spectra extraction by MCR-ALS.

# 2 Materials and Methods

## *2.1 Sample Preparation and Instrumentation*

A human serum sample from a healthy volunteer was extracted and derivatized similarly as described by [12]. Analysis were performed using a GCxGC-TOF/MS Pegasus 4D (Leco, St. Joseph, MI, USA) using a Rtx-5 (5 % diphenyl/95 % PDMS, 10 m × 0.18 mm × 0.2 μm) as a primary column, and a Rxi-17Sil MS (50 % diphenyl/50 % PDMS, 1 m × 0.15 mm ID × 0.15μm) as a secondary column, both from Restek, Bellefonte, PA, USA. The split/splitless inlet was set at 250 °C, operating in split mode at a 20:1 split rate, using He of 90.9995 % purity as a carrier gas at a constant flow of 1 ml/min. The primary oven was set at an initial temperature of 50 °C, rinsed at 20 °C/min to 285 °C, and held for 5 min. The secondary oven was set at 15 °C above the primary oven, and the modulator was set at 15 °C above the primary oven. Modulation period was of 4 s, with a hot pulse time of 0.4 s. Transfer line temperature was 250 °C. The ionization was done by electronic impact, and MS operated at an acquisition rate of 100 spectra/sec, acquiring from 25 to 500 m/z. The two-dimensional GCxGC-MS chromatogram plot of the serum sample is shown in Fig. 1.

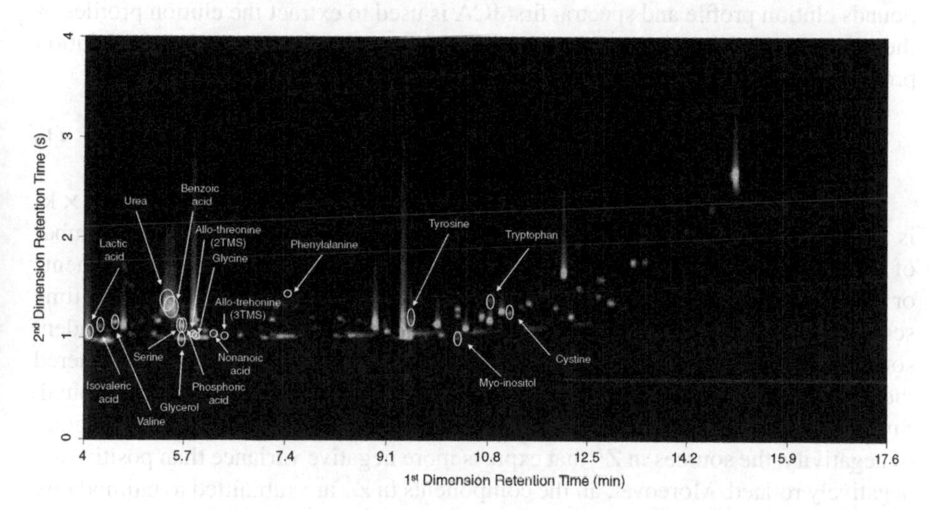


**Fig. 1** GCxGC-TOF/MS contour plot cromatogram of the human serum sample. All the identified metabolites are indicated

## *2.2 Data Processing Methods*

The GCxGC-MS samples were processed by analyzing each modulation cycle separately. A principal component analysis PCA was used to automatically determine the number of factors or components to initialize both MCR and OSD. For this study, the ICA algorithm used is the joint approximate diagonalization of eigenvalues (JADE) [13]. After being analyzed through GCxGC-MS, the serum sample was characterized by a supervised identification of certain compounds that may be found in human serum. Some metabolites were validated by analyzing their standards through GCxGC-MS. After the sample was characterized, the goal was to use OSD and MCR to automatically extract the most pure spectra for each compound. The spectra determined by OSD and MCR were compared using the cosine dot product [14] against the Golm Metabolome Database (GMD) spectra database [12]. The results include a comparative study of the identification score between MCR-ALS and OSD.

### *2.3 Spectra Extraction of GCxGC-MS Mixtures Using Orthogonal Signal Deconvolution*

Orthogonal signal deconvolution is a combination of independent component analysis and principal component analysis. As GCxGC-MS data is a combination of compounds elution profile and spectra, first ICA is used to extract the elution profiles of the compounds. After that, a PCA is used to determine the spectra for each elution profile recovered by ICA. ICA is mathematically expressed as:

$$X = AZ^T \tag{1}$$

where X ($N \times M$) is the matrix containing the mixture of compounds, A ($N \times k$) is the mixing matrix and $Z^T$ ($k \times M$) is the source matrix. N and M are the number of rows and columns of the data matrix X, and k denotes the number of components or compounds in the model. Each row in X holds the spectra for each retention time scan in each column. ICA decomposes the data matrix by finding the independent sources contained in X. In OSD, the elution profile of the compounds are considered the independent sources and thus $Z^T$ holds the elution profile for each compound. Since the elution profiles determined by ICA may be affected by the ICA ambiguity of negativity, the sources in $Z^T$ that express more negative variance than positive are negatively rotated. Moreover, all the components in $Z^T$ are submitted to unimodality constraint [15] to force one local maxima per source.

After that, an $X'_j$ sub-data matrix is determined for each compound $j$ in $Z^T$. This sub-data matrix comprises only the data from X in which the compound profile in $Z_j^T$ is non-zero. Therefore, the elution profile in $Z^T$ is used as a mask to suppress the surrounding data non-related to the compound. A PCA is performed over the sub-data matrix to determine the spectra associated to each compound. PCA can be mathematically expressed as:

$$X'_j = YW^T \tag{2}$$

where $X'(N \times M)$ is the sub-data matrix to decompose, $Y(N \times M)$ is the score matrix or eigenvalues and $W(M \times M)$ is the loading or eigenvectors matrix. For each compound profile, the PCA decorrelates the information of the sub-data matrix and decomposes it into a matrix $W^T$ (Eq. 3) which is a set of orthogonal spectra and a matrix Y which is associated to the elution profile for each spectra in $W^T$. The matrix $W^T$ holds the spectra of the compound of interest together with the spectra of the different sources of noise - such as co-eluted substances or biological matrix interference -. To determine which spectra is related to the compound of interest we compute the correlation between the profile of the compound in ${Z^T}_j$ and the information of the profiles determined by the PCA in Y. The component with the highest absolute correlation is the candidate spectra for the compound of interest.

ICA has a second ambiguity related to variance (energy) indetermination, which means that the energy of the recovered compound profiles do not correspond to the real energy of that component. To overcome that, a least squares regression is performed with the estimated sources hold in $Z^T$ against the base ion chromatogram of the matrix X. The base ion chromatogram or BIC is determined by representing the maximum *m/z* value for each point in the chromatogram.

## 3 Results and Discussion

The analysis of the human serum sample using OSD and MCR-ALS lead to the correct identification of the 16 metabolites. From all of them, only allo-threonine appeared in different trimethylsilyl (TMS) derivatives, so a total of 17 compounds were identified. The identification score for each metabolite and method is shown in Table 1. The identification score is the cosine dot product between the extracted and the reference spectra of the Golm Metabolome Database. Quantitative identification score differences appear when comparing both methods: OSD performs a more pure extraction of the spectra in the case of isovaleric acid, urea, glycerol, phosphoric acid, allo-threonine (2TMS) and (3TMS), nonanoic acid and phenylalanine, whereas MCR extracts a more pure spectra in the case of benzoic acid and serine. The overall identification results are shown in the comparative boxplot in Fig. 2 (a), where OSD shows a greater accuracy for the sample given. An example of the spectra extracted after the processing the sample with OSD is shown in Fig. 3.

As can be appreciated from the given results, the identification capability of both methods are similar since their differences are subtle. This is due to the fact that OSD is designed to operate in cases with higher noise contamination, as in the case of GC-MS. As mentioned before, one of the most important factors that difficulties the identification is coelution. In these cases, the spectra of each compound has to be separated from the possible coeluted compounds or substances. Since one of the differential characteristics of GCxGC-MS is its reduction of the coelution problem, the spectra of the compounds analyzed is more pure and less affected by noise,

**Table 1** Identification score results for the human serum sample. The highest score for each metabolite is highlighted in bold

| Rt (min) | Compound Name | OSD | MCR |
|---|---|---|---|
| 4.084 | Lactic acid (2TMS)[a] | 99.15 | **99.37** |
| 4.486 | Isovaleric acid (1TMS)[a] | **96.25** | 94.54 |
| 4.285 | Valine (1TMS) | 99.89 | **99.91** |
| 5.355 | Urea (2TMS)[a] | **96.94** | 95.88 |
| 5.488 | Benzoic acid (1TMS)[a] | 96.35 | **97.36** |
| 5.552 | Serine (2TMS) | 95.18 | **97.25** |
| 5.683 | Glycerol (3TMS)[a] | **96.60** | 95.82 |
| 5.686 | Phosphoric acid (3TMS)[a] | **96.01** | 95.84 |
| 5.818 | Threonine, allo- (2TMS)[a] | **82.41** | 81.80 |
| 5.884 | Glycine (3TMS)[a] | **97.53** | 97.50 |
| 6.218 | Nonanoic acid (1TMS) | **97.00** | 93.49 |
| 6.417 | Threonine, allo- (3TMS) | **95.32** | 92.28 |
| 7.424 | Phenylalanine (1TMS) | **98.93** | 97.68 |
| 9.553 | Tyrosine (3TMS) | **98.86** | 98.54 |
| 10.35 | Inositol, myo- (6TMS) | 96.73 | **96.83** |
| 10.89 | Tryptophan (3TMS) | **99.31** | 99.19 |
| 11.22 | Cystine (4TMS) | **96.96** | 96.37 |

[a]The marked metabolites were identified by analyzing their standards through GCxGC-MS

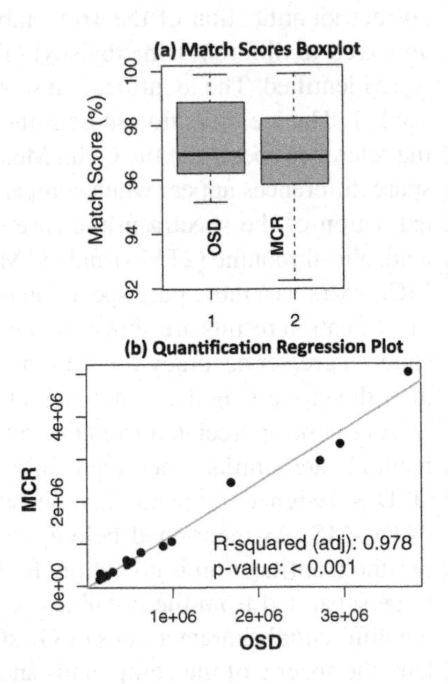


**Fig. 2** (a) Match score boxplot between all the OSD and MCR identification results. (b) Regression plot between the area of the resolved profiles between OSD and MCR

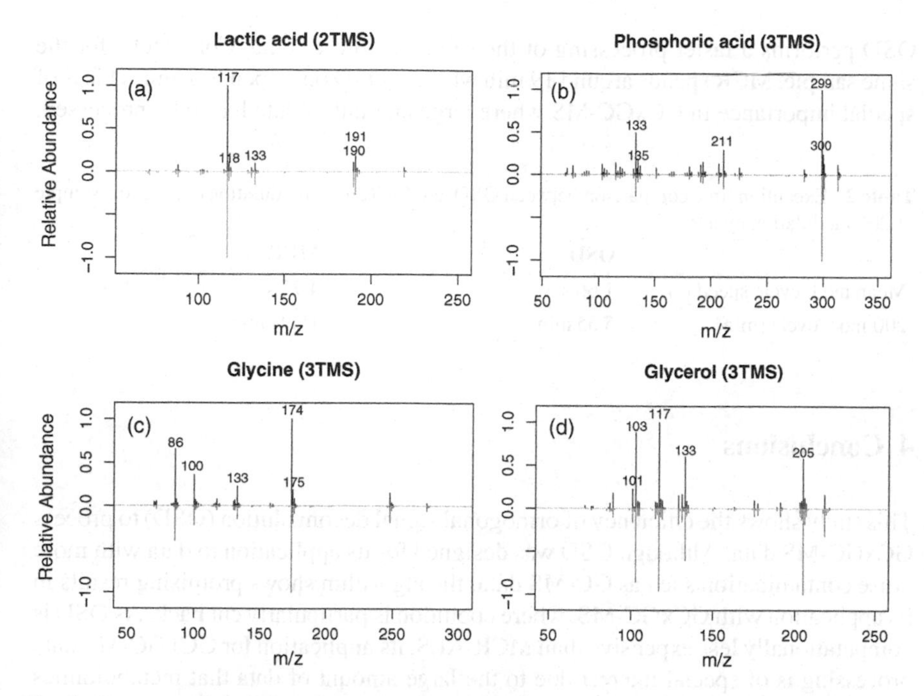


**Fig. 3** Representation of the extracted spectra (black) by OSD and the reference GMD spectra (color) in the human serum sample. Reference spectra are shown negatively rotated in the same axis for a better visual appreciation (Color figure online)

and therefore, both methods OSD and MCR determine similar identification match scores. Despite that, this shows how OSD is not overfitted and can be applied for the processing of samples with less noise interference.

For each resolved compound, both OSD and MCR determine the elution profile of each compound. The area of the most intense ion (*m/z*) of each compound is used to determine the relative concentration. Figure 2 (b) shows the regression plot of the area determined by the two methods along with the adjusted R-value and p-value. These results show that OSD is in a good agreement with the quantification given by MCR.

Finally, the computational cost and therefore the processing speed of both methods was compared. To perform that, the same modulation cycles were analyzed with the same initial conditions to initialize both algorithms. The mean processing speed for 1 modulation cycle was determined for both methods from the processing of a total of 200 modulation cycles. Those 200 modulation cycles are considered to be a representative size of an standard metabolomics sample. Data was processed using a 2.4 GHz Intel Core 2 Duo processor with 4 GB of 1067 MHz DDR3 of RAM memory. The differences of the computational cost between OSD and MCR are detailed in Table 2. The table shows the mean processing speed for one modulation cycle and the speed for 200 modulation cycles processing between methods.

OSD performs a faster processing of the same amount of data. Concretely, for the same sample, MCR spends around 19 min whereas OSD only spends 6 min. This is of special importance in GCxGC-MS where large amount of data has to be processed.

**Table 2** Execution time comparison between OSD and MCR for 1 modulation cycle and a sample of 200 modulation cycles

| | **OSD** | **MCR** |
|---|---|---|
| Mean mod. cycle speed (s) | 1.66 s | 4.77 s |
| 200 mod. cycle (min) | 5.55 min | 15.91 min |

# 4 Conclusions

This study shows the efficiency of orthogonal signal deconvolution (OSD) to process GCxGC-MS data. Although OSD was designed for its application to data with more noise contamination such as GC-MS data, the algorithm shows promising results in its application with GCxGC-MS, where coelution is particularly cut back. As OSD is computationally less expensive than MCR-ALS, its application for GCxGC-MS data processing is of special interest due to the large amount of data that metabolomics experiments generate with this analytical platform.

**Acknowledgments** The authors want to thank the Center for Omics Sciences, Tarragona, for providing the samples and to Dr. R. Ras and Mrs. S. Mariné for the sample preparation and scientific advising.

# References

1. Zhang, A., Sun, H., Wang, X.: Serum metabolomics as a novel diagnostic approach for disease: a systematic review. Anal. Bioanal. Chem. **404**(4), 1239–1245 (2012)
2. Seeley, J.V., Seeley, S.K.: Multidimensional gas chromatography: fundamental advances and new applications. Anal. Chem. **85**(2), 557–578 (2012)
3. Mondello, L., Tranchida, P.Q., Dugo, P., Dugo, G.: Comprehensive two-dimensional gas chromatography-mass spectrometry: a review. Mass Spectrom. Rev. **27**(2), 101–124 (2008)
4. Matos, J.T.V., Duarte, R.M.B.O., Duarte, A.C.: Trends in data processing of comprehensive two-dimensional chromatography: state of the art. J. Chromatogr. B **910**, 31–45 (2012)
5. Faber, N.K.M., Bro, R., Hopke, P.K.: Recent developments in CANDECOMP/PARAFAC algorithms: a critical review. Chemometr. Intell. Lab. Syst. **65**(1), 119–137 (2003)
6. van Stokkum, I.H.M., Mullen, K.M., Mihaleva, V.V.: Global analysis of multiple gas chromatographymass spectrometry (GC/MS) data sets: a method for resolution of co-eluting components with comparison to MCR-ALS. Chemometr. Intell. Lab. Syst. **95**(2), 150–163 (2009)
7. Roberts, S., Everson, R.: Independent Component Analysis: Principles and Practice. Cambridge University Press (2001)
8. Wang, G., Cai, W., Shao, X.: A primary study on resolution of overlapping GC-MS signal using mean-field approach independent component analysis. Chemometr. Intell. Lab. Syst. **82**(12), 137–144 (2006)

9. Liu, Z., Cai, W., Shao, X.: Sequential extraction of mass spectra and chromatographic profiles from overlapping gas chromatographymass spectroscopy signals. J. Chromatogr. A **1190**(12), 358–364 (2008)
10. Shao, X., Liu, Z., Cai, W.: Resolving multi-component overlapping GC-MS signals by immune algorithms. TrAC Trends Anal. Chem. **28**(11), 1312–1321 (2009)
11. Domingo-Almenara, X., Perera-Lluna, A., Ramirez, N., Canellas, N., Correig, X., Brezmes, J.: Compound deconvolution and identification in GC-MS-based metabolomics by blind source separation and principal component analysis. Article in press
12. Hummel, J., Strehmel, N., Selbig, J., Walther, D., Kopka, J.: Decision tree supported substructure prediction of metabolites from GC-MS profiles. Metabolomics **6**(2), 322–333 (2010)
13. Cardoso, J.F., Souloumiac, A.: Blind beamforming for non-gaussian signals. IEE Proc. F Radar Signal Proc. **140**(6), 362–370 (1993)
14. Wan, K.X., Vidavsky, I., Gross, M.L.: Comparing similar spectra: from similarity index to spectral contrast angle. J. Am. Soc. Mass Spectrom. **13**(1), 85–88 (2002)
15. de Juan, A., Vander Heyden, Y., Tauler, R., Massart, D.L.: Assessment of new constraints applied to the alternating least squares method. Anal. Chim. Acta **346**(3), 307–318 (1997)

# Dolphin 1D: Improving Automation of Targeted Metabolomics in Multi-matrix Datasets of $^{1}$H-NMR Spectra

**Josep Gómez, Maria Vinaixa, Miguel A. Rodríguez, Reza M. Salek, Xavier Correig and Nicolau Cañellas**

**Abstract** Nuclear magnetic resonance (NMR) is one of the main tools applied in the field of metabolomics. Extracting all the valuable information from large datasets of $^{1}$H-NMR spectra is a huge challenge for high throughput metabolomics analysis. The tools that currently exist to improve signal assignment and metabolite quantification do not have the versatility of allowing the quantification of unknown signals or choosing different quantification approaches in the same analysis. Moreover, graphical features and informative outputs are needed in order to be aware of the reliability of the final results in a field where position shifting, baseline masking and signal overlap may produce errors between samples. Here we present a software package called Dolphin 1D, which aim is to improve targeted metabolite analysis in large datasets of $^{1}$H-NMR by combining user interactivity with automatic algorithms. Its performance has been tested on a multi-matrix set composed by total serum, urine, liver aqueous extracts and brain aqueous extracts of rat. Our strategy pretends to offer a useful solution for every kind of matrix, avoiding black-box processes and subjectivities user-user in automatic signal quantification.

**Keywords** Metabolite targeted analysis · $^{1}$H-NMR · Metabolomics tool · Multi-matrix profiling

J. Gómez (✉) · M. Vinaixa · M.A. Rodríguez · X. Correig · N. Cañellas
Metabolomics Platform, IISPV, Universitat Rovira I Virgili, Campus Sescelades, Carretera de Valls, s/n, 43007 Tarragona, Catalonia, Spain
e-mail: josep.gomez@urv.cat

J. Gómez · M. Vinaixa · M.A. Rodriguez · X. Correig · N. Cañellas
CIBERDEM, Spanish Biomedical Research Centre in Diabetes and Associated Metabolic Disorders, Bonanova 69, 6th Floor, 08017 Barcelona, Catalonia, Spain

R.M. Salek
European Bioinformatics Institute (EMBL-EBI) European Molecular Biology Laboratory, Wellcome Trust Genome Campus Hinxton, CB10 1SD Cambridge, UK



R. Overbeek et al. (eds.), *9th International Conference on Practical Applications of Computational Biology and Bioinformatics*, Advances in Intelligent Systems and Computing 375, DOI 10.1007/978-3-319-19776-0_7

# 1 Introduction

Metabolomics has positioned as one of the most important techniques in the fields of molecular biology and medicine, especially for the study of metabolic-related diseases [1]. The analysis of the metabolites present in a sample can provide relevant information about the molecular processes that are occurring in a certain moment, and this molecular snap shot is very useful in order to discriminate between groups and find metabolic patterns in biological and clinical studies. In that context, nuclear magnetic resonance (NMR) spectroscopy is one of the main tools for the analysis of biofluids and tissue extracts because of its rapid, non-destructive and quantitative features [2, 3].

$^{1}$H-NMR spectra may contain hundreds of signals coming from a variable number of metabolites, but not all of them can be identified and quantified using only 1-dimensional spectra. The final number of detectable metabolites depends on the kind of biological matrix, the organism, the sample preparation and the experimental design. All this variability makes extremely difficult the automation of metabolite profiling due to signal overlapping, position shifts and baseline masking. Nowadays, the reliability and time consumption of metabolite profiling in $^{1}$H-NMR samples depends directly on user experience.

Each metabolite contributes to the final spectra with one or several signals, and, due to the issues mentioned previously, finding all the resonances of a set of target metabolites in $^{1}$H-NMR spectra is almost an impossible task. So, metabolite assignment and quantification is usually done using just one or a subset of its signals. However, a set of signals easily identifiable or/and quantifiable in one experiment may be useless in other experiments, and similar signals of different metabolites may produce false positives and false negatives when fully automated identification processes are used. Agreements between semi automatic assistance and user-interactivity may help to avoid errors and extract the maximum information possible from $^{1}$H-NMR.

While metabolite identification is the main bottleneck in NMR-metabolomics, an automatic quantification along large datasets of those previously identified metabolites is also a challenge. The most common methods for metabolite quantification in $^{1}$H-NMR spectra are bucket integration and line-shape fitting. Bucket integration is easier and faster than line-shape fitting, and it can be very useful for isolated signals but it can severely fail in regions where signal overlap is present [4–6]. Line-shape fitting is more complicated and time-consuming but essential for accurate quantifications in congested regions [7–13].

In this paper we perform a metabolite profile analysis in a multi matrix set of total serum, urine, liver aqueous extracts and brain aqueous extracts of rats using a software tool called Dolphin1D which aims to improve the targeted metabolite profile analysis in large sets of $^{1}$H-NMR spectra. This tool combines an easy and intuitive user interaction with assisted metabolite assignment and different methods of automatic quantification in order to obtain reliable results for posterior statistical analysis and optimize processing times.

## 2 Materials and Methods

### *2.1 Dataset Samples*

The dataset analyzed in this study is conformed by 168 spectra of different biofluids and tissues of rat. The total number of samples is divided in 36 of total serum, 36 of liver aqueous extracts, 36 of brain aqueous extracts and 60 of urine. The animal handling and the sample preparation of the liver aqueous extracts was performed as detailed previously in [14]. Brain aqueous extracts were obtained as described previously by [15] . Urine was collected in sterile microfuge tubes according to the method described by [16]. During sampling, urine was kept on ice and was stored after dilution with sterile saline at −80 °C until processing. For total serum samples, approximately 650 μl of blood were collected by retro-orbital bleeding just before dissection. The blood samples were incubated at 37 °C for 10 min and centrifuged for 10 min at 13100 g at 4 °C. The supernatant was collected, frozen immediately in liquid N, and was stored at −80 °C.

### *2.2 $^{1}$H-NMR Measurements*

For urine, liver aqueous extracts and brain aqueous extracts a 1D nuclear Overhauser effect spectroscopy with a spoil gradient (NOESY) was used to record 1D $^{1}$H-NMR spectra using a 600.2 MHz frequency Avance III 600 Bruker spectrometer (Bruker, Germany) equipped with an inverse TCI 5 mm cryoprobe. A total of 256 transients were collected across a 12 kHz spectral width at 300 K into 64 k data points, and exponential line broadening of 0.3 Hz was applied before Fourier transformation. A recycling delay time of 8 s was applied between scans to ensure correct quantification. $^{1}$H-NMR spectra of serum were performed using the Carr-Purcell-Meiboom-Gill sequence (cpmg spin-spin T2 relaxation filter) with a total time filter of 410 ms that attenuate the signals of macromolecules to a residual level. The spectral width was 20 ppm, and a total of 64 transients were collected into 64 k data points for each cpmg spectrum.

## 2.3 Package Overview

**Operating mode:** Dolphin1D works under a very simple and intuitive user-friendly GUI and is able to quantify a flexible set of target signals using easily editable libraries and Region Of Interest (ROI) patterns. A ROI is a spectral window containing relevant signals for metabolite quantification. The tool works with shimmed and phase corrected Bruker spectra as input, and gives the user several options to import the data (it in terms of alignment, region suppressions and normalization). When the whole dataset has been imported, the user can choose between a fully automatic high-throughput analysis and a manual supervised analysis. We recommend a manual supervised analysis the first time you work on a new dataset, because it allows the user to verify graphically how the patterns fit the new data. This mode allows the user to edit some parameters to adjust the final pattern to a new experiment before being applied to the whole dataset. For following analysis of previously analyzed data we recommend executing the fully automatic high-throughput analysis directly (see Fig. 1). All the functions are programmed under the matrix calculation platform MATLAB (ver. 7.5.0; The Mathworks, Inc., Natick, MA, USA).

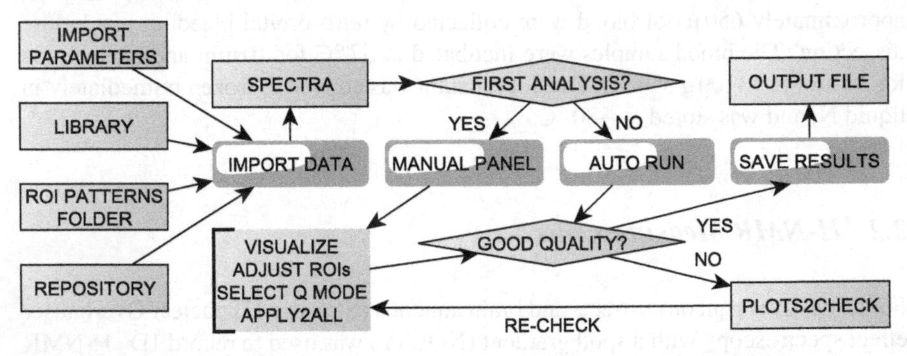


**Fig. 1** Here we have a graphical overview of the software flowchart. As shown here, the process starts by introducing 4 input files, all of them with a determined format and easily editable by the user. Once the spectra were imported, the Manual Panel allows the user to visualize, adjust and test different quantification modes with the ROIs. All the quantification iterations that do not pass a quality threshold based on the fitting error and the signal to noise ratio will be saved in a folder, giving the chance to the user to discard or re-adjust the fitting parameters in order to finally obtain the most reliable results

**Features adjustment and ROI testing through the manual panel** The tool facilitates metabolite assignations through signal suggestions in the Manual Panel. These signal suggestions come from an internal curated database built from public databases such as the Human Metabolome Database (HMDB) [17], the Birmingham Metabolite Library [18] and the BioMagResBank (BMRB) [19]. The curation is based on the selection of the most useful signals of the most relevant metabolites which frequently appear in 1H-NMR spectra of biofluids and tissues according to recent literature.

The user can expand and reduce the library and the number of ROI pattern files, and adjust the features within the ROI patterns to the experiment requirements. In the Manual Panel, the user has the option to plot a single spectrum, all spectra or the average spectrum of a ROI, which is very useful to check graphically the ROI behavior between samples. Figure 2 shows an example of a ROI testing using the Manual Panel.

Once all signals for a ROI pattern have been chosen the user can try and select which quantification mode is the most appropriate in each case before running it along the whole dataset samples. The 'Bucket Integration' mode is very useful for those regions that contain isolated and pure (without any baseline) signals because the computation time is severely reduced while the quantification remains accurate. The 'Baseline Fitting' mode allows to deconvolve targeted signals in regions where baseline or broad signals are affecting the final shape of the region, it will take more computation time but is the optimal solution in those cases. Finally, the 'Clean Fitting' mode is able to quantify accurately overlapping signals in regions where neither baseline nor broad signals are affecting.

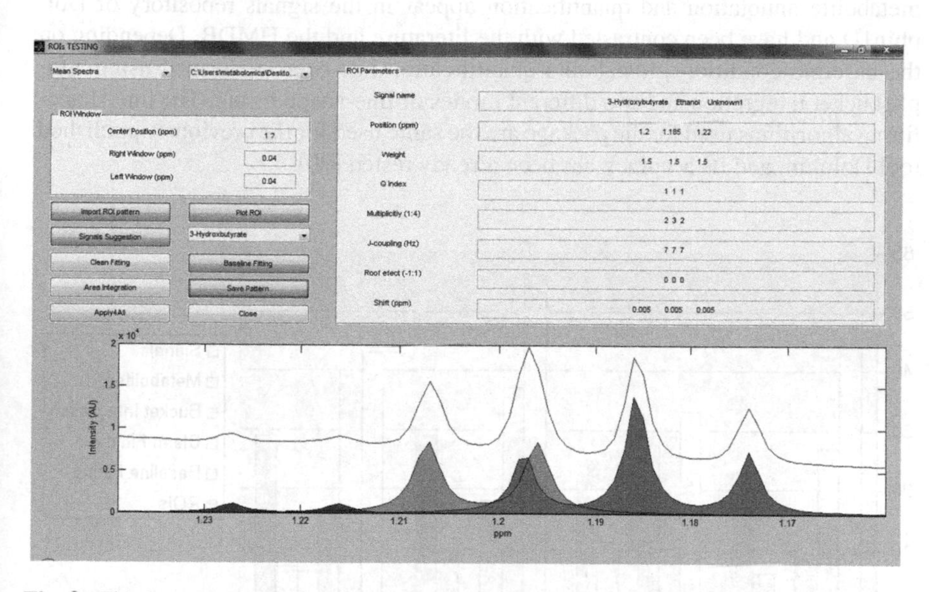


**Fig. 2** This image corresponds to a ROI pattern test on the liver aqueous extracts. We show a baseline fitting quantification of 3 signals assigned to Unknown 1 (*red*), 3-Hydroxybutyrate (pink) and Ethanol (green) using the signals suggestion tool. We used the average spectrum to adjust the fitting parameters before being applied to the whole dataset. All the fitting parameters can be modified through the ROI parameters box or in the ROI pattern file. We recommend changing the fitting parameters in the ROI pattern file only if the user wants to keep the changes as the new default version (Color figure online)

**Automatic high-throughput quantification analysis of the targeted metabolites**
The user can apply a high-throughput quantification analysis of only one of the ROIs

or apply it for all the ROIs saved in the ROI patterns folder. Whatever the case, all those iterations that do not pass a quantification quality threshold will be saved in a folder called 'Plots2Check'. During the fitting process, the algorithm adjusts not only the intensity, but also the position and the spectral width of the target signals by giving them a margin from the starting values adjusted by the user. When the analysis is finished, the user can re-run all those quantifications saved in the folder 'Plots2Check' using the Manual Panel if it is necessary. Finally, the user can save the results in an excel file.

## 3 Results and Discussion

We performed a metabolite profile analysis in a total set of 168 spectra of 4 different biological matrices of rat. Figure 3 shows the number of spectra, metabolites, ROI patterns and quantification modes used in each matrix. The signals used for metabolite annotation and quantification appear in the signals repository of Dolphin1D and have been contrasted with the literature and the HMDB. Depending on the baseline conditions, metabolite quantification has been performed using simple bucket integration and two different modes of line-shape fitting. The line-shape-fitting algorithms used for the package are the same used for the previously published tool Dolphin, and its accuracy has been already tested [20].

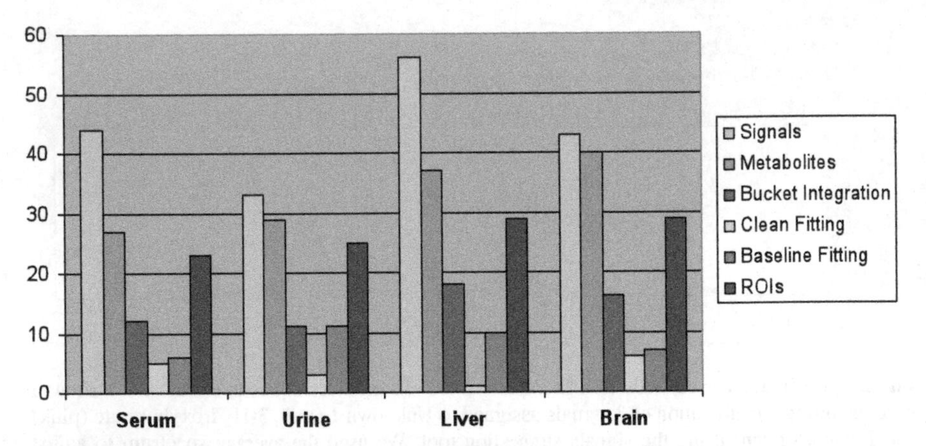


**Fig. 3** As shown in this figure, the number of ROIs is minor than the number of metabolites, and the number of metabolites is minor than the number of signals used to perform the analysis. The reason is that usually, for quantifying metabolites using line-shape fitting methods, a single ROI presents more than one signal, and not all those signals can be easily assigned to a metabolite. Anyway, the user is who ultimately decides whether quantify and/or consider unknown signals for the posterior statistical steps. The most used quantification method has been bucket integration which reduces significatively the computation time, but the number of metabolites quantified using this approach is in all cases minor than 50%

Dolphin1D aim is to quantify a flexible set of target signals using only $^1$H-NMR spectra and avoiding black-box processes. By default, quantification is given in arbitrary units and it corresponds to the fitted area of each signal. This allows the discovery of metabolite markers and the discrimination of metabolic patterns between groups. Depending on the experiment requirements, those signals related to a known metabolite resonance can be converted to absolute concentration units by quantifying a reference standard compound such as TSP or Eretic and applying the Serkova equation [21]. The number of signals can be reduced or extended depending on the user criteria, and unknown signals, artifacts and solvents can also be quantified. Dolphin1D rises as the versatile alternative of its predecessor Dolphin, whose dependence of 2-dimensional data is a clear limitation for being applied in large datasets in terms of spectroscopy acquisition times. The new package maintains the automation in terms of quantification along samples but introduces the option of user-editing ROI patterns, 'Bucket Integration' as quantification method and visual tools for tracking the process.

## 4 Conclusions

This tool has been created as an evolution of Dolphin in terms of usefulness and versatility for targeted metabolite profiling in $^1$H-NMR samples. While new algorithms need to be developed in order to improve automation in metabolite identification and quantification, a package that combines user interaction with automatic assistance can be an optimal solution in order to obtain the most information possible from $^1$H-NMR spectra in large datasets. The package has the benefits of fully automatic algorithms for quantification purposes and also the benefits of manual parameters edition and graphical checking for users to be assured their final results are reliable. The package is not matrix dependent and has default libraries and ROI patterns for different biofluids and tissues. A detailed manual of how to use it is now under construction.

## References


1. Nicholson, J.K., Lindon, J.C., Holmes, E.: 'Metabonomics': understanding the metabolic responses of living systems to pathophysiological stimuli via multivariate statistical analysis of biological NMR spectroscopic data. Xenobiotica Fate Foreign Compd. Biol. Syst. **29**(11), 1181–1189 (1999)
2. Chen, H.W., Pan, Z.Z., Talaty, N., Raftery, D., Cooks, R.G.: Combining desorption electrospray ionization mass spectrometry and nuclear magnetic resonance for differential metabolomics without sample preparation. Rapid Commun. Mass Spectrom. **20**, 1577–1584 (2006)

3. Beckonert, O., Keun, H.C., Ebbels, T.M.D., Bundy, J.G., Holmes, E., Lindon, J.C., Nicholson, J.K.: Metabolic profiling, metabolomic and metabonomic procedures for NMR spectroscopy of urine, plasma, serum and tissue extracts. Nat. Protoc. **2**, 2692–2703 (2007)
4. Alves, A.C., Rantalainen, M., Holmes, E., Nicholson, J.K., Ebbels, T.M.D.: Analytic properties of statistical total correlation spectroscopy based information recovery in H-1 NMR metabolic data sets. Anal. Chem. **81**, 2075–2084 (2009)
5. Anderson, P.E., Reo, N.V., DelRaso, N.J., Doom, T.E., Raymer, M.L.: Gaussian binning: a new kernel-based method for processing NMR spectroscopic data for metabolomics. Metabolomics **4**, 261–272 (2008)
6. Jacob, D., Deborde, C., Moing, A.: An efficient spectra processing method for metabolite identification from H-1-NMR metabolomics data. Anal. Bioanal. Chem. **405**, 5049–5061 (2013)
7. Eads, C.D., Furnish, C.M., Noda, I., Juhlin, K.D., Cooper, D.A., Morrall, S.W.: Molecular factor analysis applied to collections of NMR spectra. Anal. Chem. **76**, 1982–1990 (2004)
8. Ochs, M.F., Stoyanova, R.S., Arias-Mendoza, F., Brown, T.R.: A new method for spectral decomposition using a bilinear Bayesian approach. J. Magn. Reson. **137**, 161–176 (1999)
9. Stoyanova, R., Nicholson, J.K., Lindon, J.C., Brown, T.R.: Sample classification based on Bayesian spectral decomposition of metabonomic NMR data sets. Anal. Chem. **76**, 3666–3674 (2004)
10. Hao, J., Astle, W., De Iorio, M., Ebbels, T.M.D.: BATMAN - an R package for the automated quantification of metabolites from nuclear magnetic resonance spectra using a Bayesian model. Bioinformatics **28**, 2088–2090 (2012)
11. Hao, J., Liebeke, M., Astle, W., De Iorio, M., Bundy, J.G., Ebbels, T.M.D.: Bayesian deconvolution and quantification of metabolites in complex 1d NMR spectra using BATMAN. Nat. Protoc. **9**, 1416–1427 (2014)
12. Laatikainen, R., Niemitz, M., Malaisse, W.J., Biesemans, M., Willem, R.: A computational strategy for the deconvolution of NMR spectra with multiplet structures and constraints: Analysis of overlapping C-13-H-2 multiplets of C-13 enriched metabolites from cell suspensions incubated in deuterated media. Magn. Reson. Med. **36**, 359–365 (1996)
13. Soininen, P., Haarala, J., Vepsalainen, J., Niemitz, M., Laatikainen, R.: Strategies for organic impurity quantification by H-1 NMR spectroscopy: constrained total-line-shape fitting. Anal. Chimica Acta **542**, 178–185 (2005)
14. Vinaixa, M., Rodriguez, M.A., Rull, A., Beltran, R., Blade, C., Brezmes, J., Canellas, N., Joven, J., Correig, X.: Metabolomic assessment of the effect of dietary cholesterol in the progressive development of fatty liver disease. J. Proteome Res. **9**, 2527–2538 (2010)
15. Domange, C., Canlet, C., Traore, A., Bielicki, G., Keller, C., Paris, A., Priymenko, N.: Orthologous metabonomic qualification of a rodent model combined with magnetic resonance imaging for an integrated evaluation of the toxicity of hypochoeris radicata. Chem. Res. Toxicol. **21**, 2082–2096 (2008)
16. Cohen, S.M., Ohnishi, T., Clark, N.M., He, J., Arnold, L.L.: Investigations of rodent urinary bladder carcinogens: collection, processing, and evaluation of urine and bladders. Toxicol. Pathol. **35**, 337–347 (2007)
17. Wishart, D.S., Jewison, T., Guo, A.C., Wilson, M., Knox, C., Liu, Y., Djoumbou, Y., Mandal, R., Aziat, F., Dong, E., Bouatra, S., Sinelnikov, I., Arndt, D., Xia, J., Liu, P., Yallou, F., Bjorndahl, T., Perez-Pineiro, R., Eisner, R., Allen, F., Neveu, V., Greiner, R., Scalbert, A.: HMDB 3.0-the human metabolome database in 2013. Nucleic Acids Res. **41**, D801–D807 (2013)
18. Ludwig, C., Easton, J.M., Lodi, A., Tiziani, S., Manzoor, S.E., Southam, A.D., Byrne, J.J., Bishop, L.M., He, S., Arvanitis, T.N., Guenther, U.L., Viant, M.R.: Birmingham metabolite library: a publicly accessible database of 1-D H-1 and 2-D H-1 J-resolved NMR spectra of authentic metabolite standards (BML-NMR). Metabolomics **8**, 8–18 (2012)
19. Ulrich, E.L., Akutsu, H., Doreleijers, J.F., Harano, Y., Ioannidis, Y.E., Lin, J., Livny, M., Mading, S., Maziuk, D., Miller, Z., Nakatani, E., Schulte, C.F., Tolmie, D.E., Wenger, R.K., Yao, H., Markley, J.L.: BioMagResBank. Nucleic Acids Res. **36**, D402–D408 (2008)

20. Gomez, J., Brezmes, J., Mallol, R., Rodriguez, M.A., Vinaixa, M., Salek, R.M., Correig, X., Canellas, N.: Dolphin: a tool for automatic targeted metabolite profiling using 1d and 2d H-1-NMR data. Anal. Bioanal. Chem. **406**, 7967–7976
21. Serkova, N., Fuller, T.F., Klawitter, J., Freise, C.E., Niemann, C.U.: H-1-NMR-based metabolic signatures of mild and severe ischemia/reperfusion injury in rat kidney transplants. Kidney International **67**(3), 1142–1151 WOS:000227013500037 (2005)

# A New Dimensionality Reduction Technique Based on HMM for Boosting Document Classification

**A. Seara Vieira, E.L. Iglesias and L. Borrajo**

**Abstract** Many classification problems, such as text classification, require the ability to handle the high dimension of a structured representation of the documents. The enormous size of the data would result in burdensome computations. Consequently, there is a strong need for reducing the quantity of handled information to develop the classification process. In this paper, we propose a dimensionality reduction technique on text datasets based on a clustering method to group documents with a simple Hidden Markov Model to represent them. We have applied the new method on the OHSUMED benchmark text corpora using the $k$-NN and SVM classifiers. The results obtained are very satisfactory and demonstrate the suitability of the proposed technique for the problem of dimensionality reduction and document classification.

**Keywords** Hidden markov model · Text classification · Dimensionality reduction · Document clustering · Similarity-based classification

## 1 Introduction

With the rapid growth of corporate and digital databases, text classification has become one of the key techniques for handling and organizing text data. Text classification is the task of automatically assigning a document set to a predefined set of classes or topics [1].

---

A.S. Vieira (✉) · E.L. Iglesias · L. Borrajo
Computer Science Dept., University of Vigo,
Escola Superior de Enxeñería Informática, Ourense, Spain
e-mail: adrseara@uvigo.es

E.L. Iglesias
e-mail: eva@uvigo.es

L. Borrajo
e-mail: lborrajo@uvigo.es



R. Overbeek et al. (eds.), *9th International Conference on Practical Applications of Computational Biology and Bioinformatics*, Advances in Intelligent Systems and Computing 375, DOI 10.1007/978-3-319-19776-0_8

The representation of a document has a strong impact on the generalization accuracy of a learning system. Documents, which are typically strings of characters, have to be transformed into a representation suitable for the learning algorithm and the classification task. The most common technique in document classification tasks is the *bag-of-words* approach [2]. In this case, every document is represented by a vector where elements describe the word frequency (number of occurrences) of a certain term. The final selection of words that will be used to represent the documents is called *feature words*.

This classical text representation technique is hindered by the practical limitations of big text corpora. *Feature reduction algorithms* can be applied to the initial feature word set in order to reduce its dimensionality. Some of these techniques can remove redundant or irrelevant features from the dataset based on statistical filtering methods such as *Information Gain* [3]. In general, the application of these algorithms can improve classification accuracy and reduce the computational cost of the whole process, although it may be too costly to compute for high-dimensional datasets.

In order to represent a dataset, the *similarity-based paradigm* can also be used [4]. In this paradigm, objects (documents, in our case) are described using pairwise (dis)similarities, i.e. distances from other objects in the dataset. In this way, documents are not limited to being represented in a feature word space, and all that is needed is a way to compute distances between documents [5]. The goal of this paradigm is then to train and test a classifier using only these relational data.

Figure 1 shows an example of a typical document representation and a similarity-based representation. Given a base corpus (Fig. 1(a)), new datasets (Fig. 1(b)) can be transformed into a similarity-based representation (Fig. 1(c)) by calculating each pairwise similarity between documents. Each document is then represented by a vector of values that indicates the distance between this document and each additional document in the original base dataset.

The main problem of the similarity-based approach is the high dimensionality of the resulting similarity space [5]. In a basic approach, the dimensionality is equal to the cardinality of the base/training corpus, which can lead to computational problems in large datasets. In addition, the way in which distances between documents are measured also influences the whole process.

Following the idea of the similarity-based representation, we propose a novel feature reduction technique called DR-HMM, which combines the X-means clustering algorithm to reduce the size of the base dataset with an HMM-based document representer to compute the distances between documents. The main idea of the proposed method is to use groups of documents as the representation base instead of using single documents, so that new documents are represented by distances to groups of documents. In order to achieve that, the clustering algorithm is applied to create the groups/clusters of documents considered the base set of the new representation. In addition, the HMM to represent a document group is based on the previously developed T-HMM [6].

## 2 Materials and Methods

### 2.1 DR-HMM: Feature Reduction Method Proposal

As a text-oriented feature reduction method, DR-HMM (Dimensionality Reduction-HMM) aims to transform the representation of an initial document set into a more useful representation. This usefulness can be measured in terms of computational cost and accuracy in an automatic classification process. In this case, the goal of the technique is to lower the dimensions of the resultant feature space and transform the representation into one more useful for classifying algorithms, such as *k*-NN or SVM. Specifically, the DR-HMM is designed to take input datasets represented in a bag-of-words approach and transform them into a customized similarity-based representation.

**(a) Base corpus /Training corpus**

| | $v_0$ | $v_1$ | $v_2$ | $v_3$ | $v_4$ | $v_5$ | Class |
|---|---|---|---|---|---|---|---|
| $d_0$ | 0 | 4.4 | 0 | 2.1 | 0 | 3.1 | R |
| $d_1$ | 0 | 3 | 1.2 | 2.9 | 0 | 0 | R |
| $d_2$ | 4.2 | 0 | 0 | 2.1 | 3.3 | 1.4 | N |
| $d_3$ | 1.5 | 4 | 0 | 0.4 | 0 | 0 | N |

**(b) Corpus to transform**

| | $v_0$ | $v_1$ | $v_2$ | $v_3$ | $v_4$ | $v_5$ | Class |
|---|---|---|---|---|---|---|---|
| $d_4$ | 0 | 0 | 4 | 0 | 0 | 2.4 | ? |
| $d_5$ | 1.1 | 2 | 0 | 3.5 | 4.8 | 0 | ? |
| $d_6$ | 3.2 | 1.2 | 0 | 0 | 0 | 0 | ? |
| $d_7$ | 0 | 4.1 | 0 | 1.1 | 0 | 3.3 | ? |

**(c) Converted corpus**

| | a_$d_0$ | a_$d_1$ | a_$d_2$ | a_$d_3$ | Class |
|---|---|---|---|---|---|
| $d_4$ | $s(d_4,d_0)$ | $s(d_4,d_1)$ | $s(d_4,d_2)$ | $s(d_4,d_3)$ | ? |
| $d_5$ | $s(d_5,d_0)$ | $s(d_5,d_1)$ | $s(d_5,d_2)$ | $s(d_5,d_3)$ | ? |
| $d_6$ | $s(d_6,d_0)$ | $s(d_6,d_1)$ | $s(d_6,d_2)$ | $s(d_6,d_3)$ | ? |
| $d_7$ | $s(d_7,d_0)$ | $s(d_7,d_1)$ | $s(d_7,d_2)$ | $s(d_7,d_3)$ | ? |

**Fig. 1** Basic similarity-based corpus transformation. (**a**) Base/Training corpus using a typical bag-of-words representation. It contains six feature words $v_i$. The *Class* attribute is binary and can take the values: Relevant (R) and Non-relevant (N) (**b**) New dataset to transform into a similarity-based representation (**c**) Dataset converted into a similarity-based representation based on the training corpus. Documents are represented in terms of distances/similarities (*s* function) to the original documents in the base corpus

The DR-HMM model is built around a training corpus that is used as the document set on which the output similarity space is based. Figure 2 shows the main workflow applied to create the DR-HMM. In the building process, two main phases can be distinguished: (i) Document clustering of the base corpus; (ii) Creation and training of the models to represent each cluster.

In order to avoid using all the instances in the training corpus as the document base in the similarity space, we apply a clustering technique to reduce the resultant feature dimensionality. Thus, instead of using single documents, document clusters are taken as the set on which the similarity space is based. In our work, the X-means algorithm [7] is chosen. This algorithm is an extension of the standard K-means technique in which the number of clusters does not need to be previously determined.

The clustering algorithm is applied per each class in the training corpus, as seen in Fig. 2. Documents are split into groups according to their class value. The clustering process is then performed for each group, producing a resultant cluster set for each class. Since the document clusters are the base of the new feature space, an additional way of computing distances between new documents and these clusters needs to be defined. This is where the HMM is integrated in the method. An HMM is created and trained with each cluster, as is shown in Fig. 2. The next section shows how the HMM model is defined and trained in order to represent the documents within a cluster.

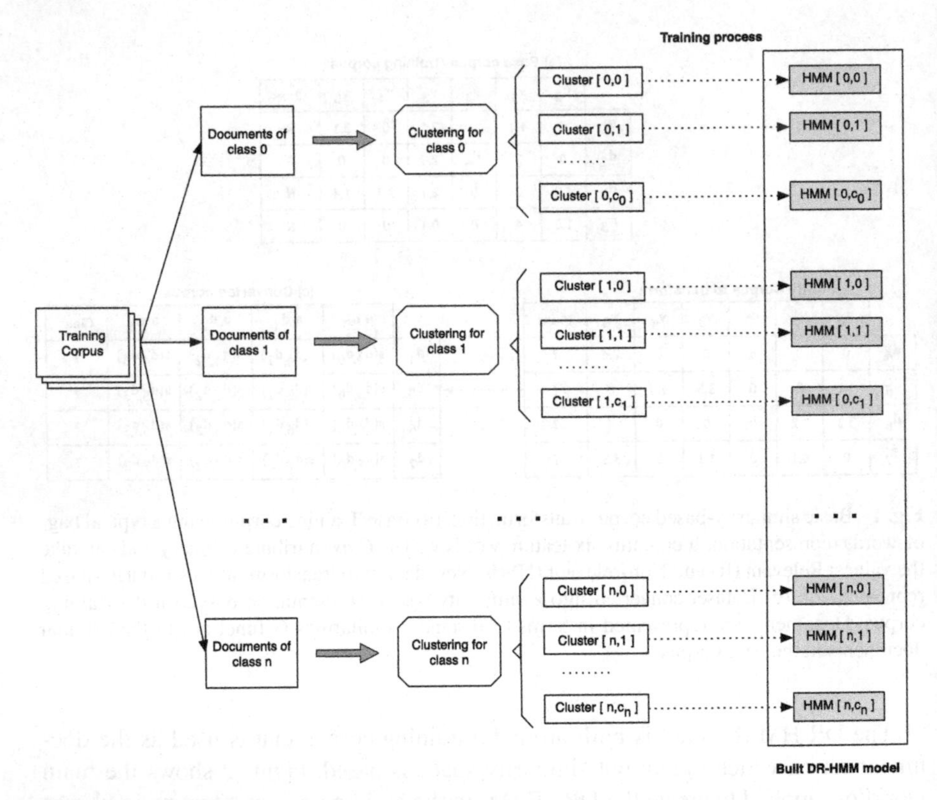


**Fig. 2** General workflow to build the DR-HMM model which uses an initial training corpus as input

## 2.2 Using HMM to Represent a Document Cluster

In a previous work, the authors have developed an HMM based document classifier called T-HMM [6]. In this model, HMMs are used to represent sets of documents. An HMM is trained per class with the documents belonging to that class . When a new document needs to be classified, the model evaluates the probability of this document being generated by each of the HMMs and outputs the class with the maximum probability value.

HMMs with the same structure as in T-HMM are used to represent document clusters. In order to do so, each HMM representing a cluster needs to be trained with the documents assigned to that cluster. The training process of the HMMs with a document set as input is the same as described in [6]. The probability distributions of the HMMs are adjusted automatically depending on the content of the documents and only two additional parameters need to be fixed to start the process: the number of stats and the generalization factor. The values of these are detailed in Sect. 2.4.

## 2.3 Data Transformation

The previous training process is applied for each cluster generated in the clustering step, since one HMM is created to represent each cluster. Once all the HMMs are trained, the DR-HMM model is considered to be finally built, and can be used to transform every single dataset into the final HMM-based similarity space.

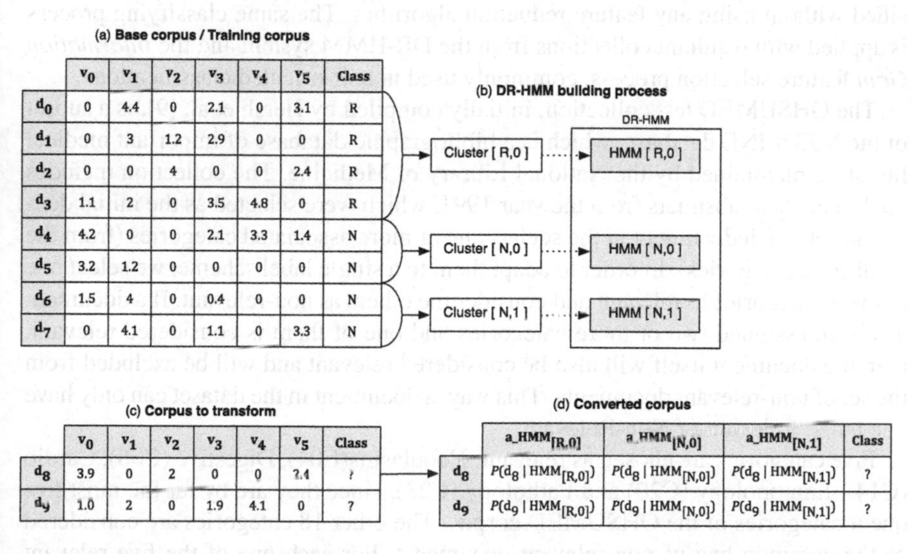


(a) Base corpus / Training corpus

| | $v_0$ | $v_1$ | $v_2$ | $v_3$ | $v_4$ | $v_5$ | Class |
|---|---|---|---|---|---|---|---|
| $d_0$ | 0 | 4.4 | 0 | 2.1 | 0 | 3.1 | R |
| $d_1$ | 0 | 3 | 1.2 | 2.9 | 0 | 0 | R |
| $d_2$ | 0 | 0 | 4 | 0 | 0 | 2.4 | R |
| $d_3$ | 1.1 | 2 | 0 | 3.5 | 4.8 | 0 | R |
| $d_4$ | 4.2 | 0 | 0 | 2.1 | 3.3 | 1.4 | N |
| $d_5$ | 3.2 | 1.2 | 0 | 0 | 0 | 0 | N |
| $d_6$ | 1.5 | 4 | 0 | 0.4 | 0 | 0 | N |
| $d_7$ | 0 | 4.1 | 0 | 1.1 | 0 | 3.3 | N |

(c) Corpus to transform

| | $v_0$ | $v_1$ | $v_2$ | $v_3$ | $v_4$ | $v_5$ | Class |
|---|---|---|---|---|---|---|---|
| $d_8$ | 3.9 | 0 | 2 | 0 | 0 | 2.4 | ? |
| $d_9$ | 1.0 | 2 | 0 | 1.9 | 4.1 | 0 | ? |

(d) Converted corpus

| | $a\_HMM_{[R,0]}$ | $a\_HMM_{[N,0]}$ | $a\_HMM_{[N,1]}$ | Class |
|---|---|---|---|---|
| $d_8$ | $P(d_8 \mid HMM_{[R,0]})$ | $P(d_8 \mid HMM_{[N,0]})$ | $P(d_8 \mid HMM_{[N,1]})$ | ? |
| $d_9$ | $P(d_9 \mid HMM_{[R,0]})$ | $P(d_9 \mid HMM_{[N,0]})$ | $P(d_9 \mid HMM_{[N,1]})$ | ? |

**Fig. 3** Example of feature transformation using a DR-HMM model. (a) Training corpus to build the DR-HMM. (b) DR-HMM building process in which three final clusters are created. (c) Dataset to be transformed (d) Resultant transformed dataset. It has one synthetical attribute per cluster

Figure 3 shows an example of the DR-HMM model building. In this case, the training corpus (Fig. 3(a)) has two possible classes for the documents, Relevant or Non-relevant. The clustering process produces a total of three clusters: two for the non-relevant class and one for the relevant class. The clusters are taken as the training sets for each HMM created (Fig. 3(b)). When a new document set (with the same feature space representation) arrives (Fig. 3(c)), the DR-HMM consisting of the three HMMs is used to transform it into the new similarity space (Fig. 3(d)).

The resultant dataset has a synthetical attribute $a\_HMM_{[x,y]}$ per trained HMM. In order to transform a document $d$, the similarities between this document and each cluster (represented by the HMMs) need to be computed. The similarities between a document and document clusters are calculated by applying the forward-backward algorithm explained in the Rabiner article [8]. In this case, the similarity between a document $d$ and a cluster represented by the HMM $\lambda_h$ is equal to the probability of $d$ being generated by $\lambda_h$. This is $P(L_d | \lambda_h)$, which is what forward-backward algorithm calculates.

## 2.4 Experiments Description

The goal of the experiments is to test the performance of the proposed feature reduction technique in the OHSUMED document corpora. The collection is described in the following section, and is used in a classification process with the SVM and $k$-NN classifiers, after applying the dimensionality reduction methods. In order to compare the effectiveness of the proposed feature reduction technique, the corpora is also classified without using any feature reduction algorithm. The same classifying process is applied with resultant collections from the DR-HMM system and the *Information Gain* feature selection process, commonly used in text-oriented classifications.

The OHSUMED test collection, initially compiled by Hersh et al. [9], is a subset of the MEDLINE database, which is a bibliographic database of important medical literature maintained by the National Library of Medicine. The collection includes 50,216 medical abstracts from the year 1991, which were selected as the initial document set. Each document in the set has one or more associated categories (from the 23 disease categories). In order to adapt them to a single label scheme, we select one of these categories as relevant and consider the others as non-relevant. If a document has been assigned two or more categories and one of them is considered relevant, then the document itself will also be considered relevant and will be excluded from the set of non-relevant documents. This way, a document in the dataset can only have one label: Relevant or Non-Relevant.

Five categories are chosen as relevant: Neoplasms(C04), Digestive (C06), Cardio (C14), Immunology (C20) and Pathology (C23), since they are by far the most frequent categories of the OHSUMED corpus. The other 18 categories are considered as the common bag of non-relevant documents. For each one of the five relevant categories, a different corpus is created in the way mentioned above, ending up with five distinct matrices.

**Evaluation Process** Initially, the document corpora need to be pre-processed. Following the bag-of-words approach, we format every document into a vector of feature words in which elements describe the word occurrence frequencies. All the different words that appear in the training corpus are candidates for feature words. In order to reduce the initial feature size, standard text pre-processing techniques are used. A predefined list of stopwords (common English words) is removed from the text, and a stemmer based on the Lovins stemmer is applied. Then, words occurring in fewer than ten documents of the entire training corpus are also removed. In addition, the tf-idf statistic is applied in order to measure the word relevance. The application of tf-idf decreases the weight of terms that occur very frequently in the collection, and increases the weight of terms that occur rarely [10]. In the final step of the pre-process phase, the initial dataset matrix is divided into two different matrices: the train matrix and the test matrix. This division is performed randomly while maintaining the class proportion in the original corpus.

Once the pre-processing phase is finished, the feature reduction process is applied to further decrease the dimensionality of the corpus. The train matrix is used as the base corpus in the following algorithms:

- The feature selection method based in Information Gain (implemented in WEKA) is used as the base feature reduction algorithm, since it was previously employed in similar text classification tasks [3, 11]. This algorithm uses a threshold to determine the size of the reduced feature set. In this case, the threshold is set to the minimum value, so that every feature word with a non-zero value of information gain is included in the resultant feature set.
- A DR-HMM system is built with the resultant corpus from the application of the Information Gain method. The clustering algorithm applied is the previously mentioned X-means technique, where the number of clusters does not need to be specified. The number of states $N$ is set to the average number of feature words with a non-zero frequency value in the cluster documents, and the $f$-factor is set to $f = 0.20$, since it provided the best results in the experiments.

The formatted train and test matrices obtained from the application of InfoGain and DR-HMM methods are used in the next classification phase. In this case, the following classification algorithms are trained to perform the document classification task:

- $k$-NN: The number of neighbours is set to $k = 3$ since it leads to the best performance of the algorithm in the tested corpus.
- Support Vector Machine (SVM): The implementation of the SVM used in this case is LIBSVM [12] and the parameters are those utilized by default in the WEKA environment, applying a RBF kernel.

In order to evaluate the effectiveness of the DR-HMM system and compare it with the Information Gain based algorithm, the complete process is executed for each method. Specifically, the entire process was executed ten times per corpus, classifier and feature reduction method.

## 3 Results and Discussion

To evaluate the effectiveness of the models, $F$-measure ($F$-meas) and Kappa Statistic, evaluation measures commonly utilized in text classification and information retrieval, are used. Table 1 shows the results achieved. The values correspond to the average value achieved in that measure for the total of 10 executions with each method. In addition, the $F$-measure value is associated with the relevant class of the documents. This measure is selected since the relevant class is the minority class in every corpus. Using this measure helps verify the differences between the models since this value is more sensitive to changes than its non-relevant counterpart.

Clearly, the new similarity representation offered by the DR-HMM leads to a large improvement of the classification performance with the $k$-NN algorithm. In the case of SVM, the values achieved with Information Gain and DR-HMM are much closer, and the DR-HMM system outperforms the Information Gain model in three of the five tested corpus.

**Table 1** The best values among classifiers for a specific measure and corpus are shown in bold

| | | $k$-NN | | | SVM | | |
|---|---|---|---|---|---|---|---|
| Corpus | Measure | Raw | InfoGain | DR-HMM | Raw | InfoGain | DR-HMM |
| Ohsumed | $F$ | 0.426 | 0.479 | **0.789** | 0.755 | 0.788 | **0.807** |
| C04 | Kappa | 0.215 | 0.378 | **0.715** | 0.690 | 0.726 | **0.745** |
| Ohsumed | $F$ | 0.096 | 0.475 | **0.624** | 0.650 | 0.761 | **0.776** |
| C06 | Kappa | 0.068 | 0.437 | **0.551** | 0.617 | 0.729 | **0.745** |
| Ohsumed | $F$ | 0.243 | 0.422 | **0.800** | 0.785 | **0.824** | 0.817 |
| C14 | Kappa | 0.143 | 0.340 | **0.735** | 0.730 | **0.771** | 0.763 |
| Ohsumed | $F$ | 0.115 | 0.422 | **0.620** | 0.561 | 0.579 | **0.696** |
| C20 | Kappa | 0.092 | 0.380 | **0.548** | 0.523 | 0.539 | **0.656** |
| Ohsumed | $F$ | 0.470 | 0.424 | **0.554** | 0.493 | **0.580** | 0.557 |
| C23 | Kappa | 0.058 | 0.200 | **0.274** | 0.346 | **0.393** | 0.348 |

## 4 Conclusions

This paper presents a novel technique for reducing the input space in order to improve the efficiency of the text classifiers. The technique utilizes a document clustering to separate data into groups, and introduces a similarity-based document representation based on a Text Hidden Markov Model. The approach is convenient for large datasets. The experimental results of the study show that the document classification accuracy obtained after the dimensionality reduction using the proposal technique is, in general, better than the original accuracy if a statistical filtering method such as Information Gain is applied.

As future work, the improvement of the underlying HMM model that measures the relevance of words can directly benefit the DR-HMM system. In addition, simplier corpus that allow the execution of more complex feature reduction techniques like PCA (Principal Component Analysis) can be added in order to compare their results.

**Acknowledgments** This work has been funded from the European Union Seventh Framework Programme [FP7/REGPOT-2012-2013.1] under grant agreement n 316265, BIOCAPS, and the "Platform of integration of intelligent techniques for analysis of biomedical information" project (TIN2013-47153-C3-3-R) from Spanish Ministry of Economy and Competitiveness.

## References

1. Sebastiani, F.: Text categorization. In: Text Mining and its Applications to Intelligence, CRM and Knowledge Management, pp. 109–129. WIT Press (2005)
2. Tsimboukakis, N., Tambouratzis, G.: Document classification system based on hmm word map. In Proceedings of the 5th International Conference on Soft Computing as Transdisciplinary Science and Technology, CSTST '08, ACM, pp. 7–12, New York, NY, USA (2008)
3. Janecek, A.G., Gansterer, W.N., Demel, M.A., Ecker, G.F.: On the relationship between feature selection and classification accuracy. JMLR Workshop Conf. Proc. **4**, 90–105 (2008)
4. Pekalska, E., Duin, R.P.W.: Dissimilarity representations allow for building good classifiers. Pattern Recogn. Lett. **23**, 943–956 (2002)
5. Bicego, M., Murino, V., Figueiredo, M.A.T.: Similarity-based classification of sequences using hidden markov models. Pattern Recogn. **37**(12), 2281–2291 (2004)
6. Seara Vieira, A., Iglesias, E.L., Borrajo, L.: T-HMM: a novel biomedical text classifier based on hidden markov models. In: 8th International Conference on Practical Applications of Computational Biology and Bioinformatics (PACBB 2014), volume 294 of Advances in Intelligent Systems and Computing, pp. 225–234. Springer International Publishing (2014)
7. Pelleg, D., Moore, A.W.: X-means: extending k-means with efficient estimation of the number of clusters. In Proceedings of the Seventeenth International Conference on Machine Learning, ICML '00, Morgan Kaufmann Publishers Inc, pp. 727–734, San Francisco, CA, USA (2000)
8. Rabiner, L.R.: Readings in speech recognition. Chapter A tutorial on hidden Markov models and selected applications in speech recognition, pp. 267–296. Morgan Kaufmann Publishers Inc., San Francisco, CA, USA (1990)
9. Hersh, W.R., Buckley, C., Leone, T.J., Hickam, D.H.: Ohsumed: an interactive retrieval evaluation and new large test collection for research. In SIGIR, pp. 192–201 (1994)
10. Baeza-Yates, R.A., Ribeiro-Neto, B.: Modern Information Retrieval. Addison-Wesley Longman (1999)
11. Caporaso, J.G., Baumgartner, W.A., Cohen, K.B., Johnson, H.L., Paquette, J., Hunter, L.: Concept recognition and the trec genomics tasks. In: Voorhees, E.M., Buckland, L.P. (eds.), TREC, volume Special Publication 500–266. National Institute of Standards and Technology (NIST) (2005)
12. Chang, C., Lin, C.: Libsvm: a library for support vector machines. ACM Trans. Intell. Syst. Technol. **2**(3):27:1–27:27 (2011)

# Diagnostic Knowledge Extraction from MedlinePlus: An Application for Infectious Diseases

**Alejandro Rodríguez-González, Marcos Martínez-Romero, Roberto Costumero, Mark D. Wilkinson and Ernestina Menasalvas-Ruiz**

**Abstract** In the creation of diagnostic decision support systems (DDSS) it is crucial to have validated and precise knowledge in order to create accurate systems. Typically, medical experts are the source of this knowledge, but it is not always possible to obtain all the desired information from them. Another valuable source could be medical books or articles describing the diagnosis of diseases managed by the DDSS, but again, it is not easy to extract this information. In this paper we present the results of our research, in which we have used Web scraping and a combination of natural language processing techniques to extract diagnostic criteria from MedlinePlus articles about infectious diseases.

**Keywords** Diagnostic knowledge · Information extraction · CDSS · DDSS · NLP

---

The erratum of this chapter can be found under DOI 10.1007/978-3-319-19776-0_16

---

A. Rodríguez-González (✉) · M.D. Wilkinson
Universidad Politécnica de Madrid – Centro de Biotecnología y Genómica de Plantas, México, México
e-mail: alejandro.rodriguezg@upm.es

M.D. Wilkinson
e-mail: mark.wilkinson@upm.es

M. Martínez-Romero
Universidad de A Coruña – Centro IMEDIR, A Coruña, Spain
e-mail: marcosmartinez@udc.es

R. Costumero · E. Menasalvas-Ruiz
Universidad Politécnica de Madrid – Centro de Biotecnología Biomédica, Madrid, Spain
e-mail: roberto.costumero@upm.es

E. Menasalvas-Ruiz
e-mail: ernestina.menasalvas@upm.es



R. Overbeek et al. (eds.), *9th International Conference on Practical Applications of Computational Biology and Bioinformatics*, Advances in Intelligent Systems and Computing 375, DOI 10.1007/978-3-319-19776-0_9

# 1 Introduction

Practitioners of medicine who seek "medical knowledge", and more specifically, "diagnostic knowledge", in a structured way face several complications. It is very difficult to find open and reliable data sources where the diagnostic criteria of diseases are published in a machine readable format making them easy to reuse.

Several authors have proposed a multitude of approaches for the extraction of medical knowledge from different sources. Some examples are Tsumoto [1], Tan et al. [2], Hahn et al. [3] or Amaral et al. [4]. However, most previous efforts are not focused on extraction of the basic clinical terms used to express the diagnostic criteria of a disease. Most of these works are focused on specific medical areas, such as radiology, or on complex, specialized tasks.

In this research, we have conceived, tested and evaluated a new way of extracting relevant medical diagnostic terms from a set of online MedlinePlus articles about infectious diseases. We have developed a prototype capable of crawling the HTML code of the Web pages in order to extract all relevant content about the diagnosis (symptoms, signs and diagnostic tests). Then, a Natural Language Processing (NLP) method based on MetaMap [5] is applied to extract relevant terms. After that, the terms provided by MetaMap are validated using a set of reference terminological resources. Once the validation has been performed the final list of relevant terms for a given disease is provided.

The reminder of the paper is organized as follows: Sect. 2 presents related work. Section 3 provides a detailed explanation about the method that has been conceived and the prototype that has been built. Section 4 presents the results of the experiments performed to test the validity of the proposed approach. Finally, Sect. 5 concludes the paper and sketches some future research directions.

# 2 Related Work

There are several sources of information which contain queryable diagnostic knowledge. However, some sources are very focused, such as Online Mendelian Inheritance in Man (OMIM) [6] and the Human Phenotype Ontology (HPO) [7], and not always contain the demanded knowledge. Ontology-like sources such as Unified Medical Language System (UMLS) [8] can provide valuable, but somewhat limited, diagnostic knowledge information, and textual databases such as Wikipedia and Freebase also contain valuable knowledge, but the reliability and completeness of their information is questionable. Finally, another well-known resource is the DiseasesDatabase,[1] however it is not possible to access its' raw data. One of the most task-specific resources, with respect to medical diagnostics, is

[1]http://www.diseasesdatabase.com/.

MedlinePlus; however there is currently no method for extracting the explicit knowledge contained within its' text.

Diagnostic knowledge extraction from medical text has been undertaken in several notable studies. Okumura et al. [9] stated that "*clinical decision support systems necessitates a diagnostic knowledge base which comprises a set of clinical findings for each disease*", and performed an analysis of the mapping between clinical vocabularies and findings in medical literature using OMIM as a knowledge source and MetaMap as the NLP tool. Another very interesting approach was undertaken by Okumura and Tatesi [10] where the authors analyze the application of MetaMap to the efficient extraction of symptomatic expressions, adding some heuristics to exploit patterns of tag sequences that frequently appear in typical symptomatic expressions.

A more general study was done by Wu et al. [11] comparing current clinical NLP systems to detect abbreviations in discharge summaries (an interesting approach with respect to our efforts here to identify abbreviated entities). In the same context, Denecke, K. [12] provides a qualitative study about the extraction of medical concepts in medical social media through NLP tools.

With the analysis of related work we can conclude that current efforts are aligned with our approach. However, our work is more focused on the extraction plus validation of the terms by means of external datasets and sources of knowledge.

# 3 Materials and Methods

The process of diagnosing a disease is mainly based on the identification of clinical observations (findings) of a patient. These findings (symptoms, signs and diagnostic tests) allow the physician to determine (or discard) a list of possible diseases or evaluate the procedures that will help in the selection of the final diagnosis [13]. Findings are also essential for researchers to examine relationships between diseases based on symptom-similarity, and to find genetic relationships between the molecular origins of diseases and their resulting phenotypes [14].

Based on the importance of these information-types, it is crucial to have automated methods capable of extracting findings and their associated diseases from unstructured data sources. In this section we present the main details of the solution we implemented to achieve this goal.

## *3.1 Information Source*

MedlinePlus is an online information service provided by the US National Library of Medicine. It provides curated health information in English and Spanish, including encyclopedic information about health and drug issues among other services. As part of this encyclopedic information, MedlinePlus provides information about a wide number of diseases. The data provided for each disease may vary, but it usually includes a description of the disease, causes, symptoms, exams and tests, and treatment.

The analysis of the HTML code of the pages provided by MedlinePlus reveals that it is predictably structured, in contrast to other well-known health web pages with diseases information such as the CDC federal agency website. This reliability of source-data structure, together with the reliability of the MedlinePlus data itself, was the determining factor in its selection as the information source for our research.

## 3.2 Architecture and General Workflow

Our proposed solution is based on the architecture depicted in Fig. 1. The general workflow of our solution is based on three separate procedures:

**Medical Text Extraction and NLP Procedures (MTENP Procedure)**

This procedure comprises the Medical Text Extraction module and the MetaMap filter and produces a list of annotated medical terms. The URL for the selected disease is sent to the Medical Text Extraction module. This module applies a Web scraping procedure to the page using the JSoup API,[2] which extracts the text of the relevant sections of the Web page. The sections considered relevant were: "Symptoms" and "Exam and Tests".

After Web scraping, the NLP procedure (MetaMap Filter) applies MetaMap to the extracted text. MetaMap restricts semantic types to "Diagnostic Procedure", "Disease or Syndrome", "Finding", "Laboratory Procedure", "Laboratory or Test Result" and "Sign or Symptom". The filter process results in a list of annotated medical terms which are relevant based on these semantic types from SNOMED-CT.

**Validation Terms Extraction Procedure (VTE Procedure)**

This procedure includes the Validation Terms Extraction module which generates a validation terminology applied to the terms produced by MetaMap within the MTENP Procedure. Given that the terms produced by MetaMap might not be correct (for example: the word "red", which is classified by MetaMap as a finding), this downstream validation of concepts by the VTE Procedure improves the overall accuracy of the approach.

The VTE Procedure is responsible for obtaining medical terms from several different sources. As potential sources of information, VTE distinguish between:

- Trusted sources: Sources under institutional maintenance or created by trusted institutions. Includes: ICD9CM, ICD10CM and MeSH.

[2]http://jsoup.org/.

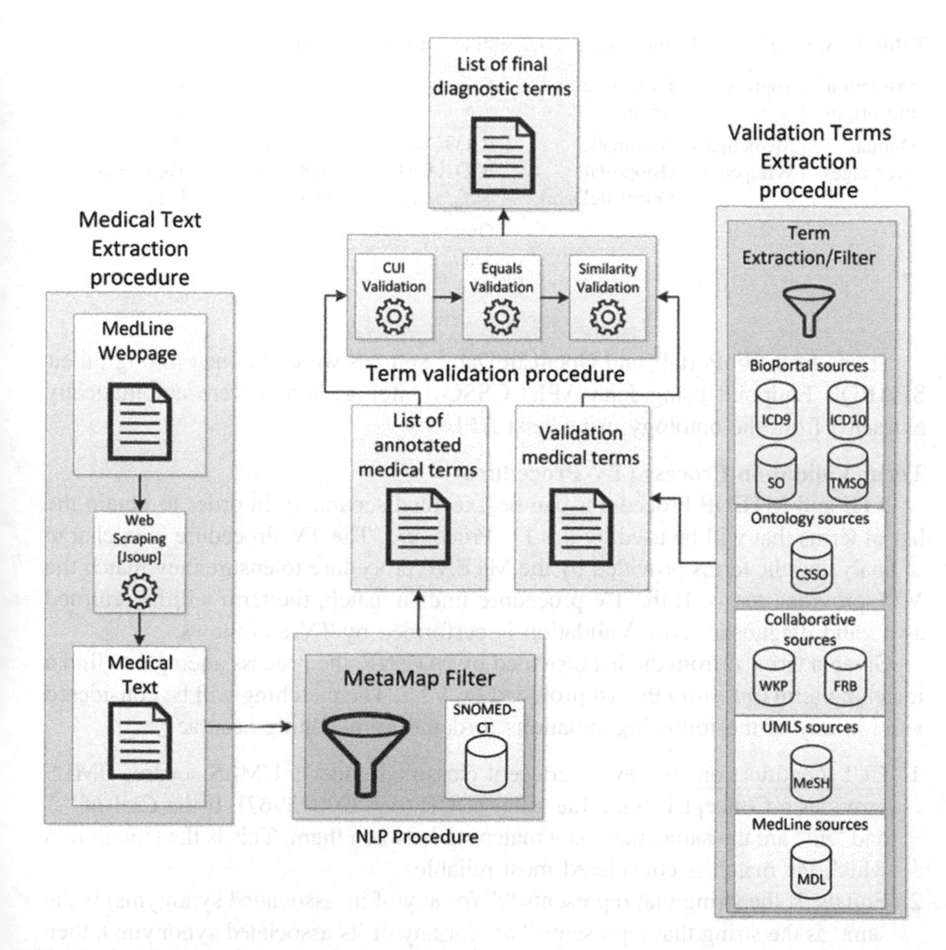


**Fig. 1** Architecture of the proposed solution.

- Research sources: Sources created as part of a research. Includes: *CCSO Signs and Symptoms Ontology*, *TM Signs and Symptoms Ontology* (TM SSO) and *Symptoms Ontology*.
- Collaborative sources: Wikipedia and Freebase.
- Other sources: Other medical webs. Includes: Medicinet[3] (Tests section).

VTE obtains the list of terms differently depending on the source (see Table 1).

[3]http://www.medicinenet.com/procedures_and_tests/article.htm.

**Table 1** Summary of validation sources and extraction method used

| Extraction and origin | Sources | Extraction and origin | Sources | Extraction and origin | Sources |
|---|---|---|---|---|---|
| Manual (webpage) | Medicinet Wikipedia | Automatic (Bioportal OpenLifeData) | ICD9CM ICD10CM Symptoms Ontology TM SSO | Automatic (Jena and MQL) | Freebase (MQL) CSSO (Jena) |

Terms from BioPortal[4] and OpenLifeData[5] sources were obtained through their SPARQL Endpoint using Jena API.[6] CSSO Ontology terms were automatically extracted from the ontology using Jena API.

**Term Validation Process (TV Procedure)**

VTE and MTENP Procedures can be executed separately in order to obtain the list of terms that will be used by the TV Procedure. The TV Procedure is in charge of analyzing the terms provided by the MTENP procedure to ensure they match the VTE-provided terms. If the TV procedure finds a match, the term will be returned as a valid diagnostic term. Validation is performed by TV as follows:

Given a term (*t*) from the list provided by MTENP, the process attempts to find a matching term (*mt*) from the list provided by VTE. The matching will be considered valid in any of the following situations (ordered by matching accuracy):

1. CUI Identification: For every concept classified under a UMLS source, UMLS provides a Concept Unique Identifier (CUI) (e.g. C0015967). If the CUI of "*t*" and "*mt*" are the same, there is a matching between them. This is the situation in which the match is considered most reliable.
2. Equals: If the string that represents "*t*" (or any of its associated synonyms) is the same as the string that represents "*mt*" (or any of its associated synonyms), then there is a match between them.
3. Similarity: A similarity score between "*t*" (and any of its synonyms) and "*mt*" (or any of its synonyms) is calculated, by means of the Levenshtein distance algorithm, with a threshold value of 0.85. We have used the implementation of the Levenshtein distance algorithm provided by the SimMetrics Java API.[7] The terms "*t*" and "*mt*" are pre-processed previously in order to maximize the possibility of finding a similarity. Pre-processing includes removal of stop words, symbols and trimming of the string among others.

If a matching is found, it is assumed that the term "*t*" is a valid diagnostic term and it is added to the final list of results.

[4]http://bioportal.bioontology.org/.

[5]http://www.openlifedata.org/.

[6]https://jena.apache.org/.

[7]http://sourceforge.net/projects/simmetrics/.

The process carried out by VTE tries to give preference to those validation terms who came, first, from trusted sources; second, from research sources; third, from collaborative sources; and finally from other sources, in order of priority.

The source code of the prototype is publicly available at GitHub.[8]

## 4 Evaluation

The approach has been tested by executing the prototype over data from 30 different infectious diseases,[9] selected manually by our researchers. Infectious diseases typically have a large number of symptoms and diagnostic tests, providing a a large variety of terms that should be extracted by our platform.

The evaluation was performed by doing a manual analysis of the results provided by our approach. For each disease, we compared: (1) the list of terms provided by our approach; with: (2) a list of terms manually extracted from the disease Web page.

True positive (TP), false positive (FP), true negative (TN) and false negative (FN) parameters were computed in order to calculate precision, recall, specificity and F1 score values. The mean values for these parameters based on the individual values for each disease are shown in Fig. 2. Detailed results for each disease are available online,[10] including the list of terms manually extracted from the disease web page, the matching with the list of terms provided by our approach and some extra information such as the primary type of source used for the matching.

The results show that our method performed well. A detailed analysis of the results, disease-by-disease, shows that there were several "false negatives because validation (a subtype of false negative – see raw results)", which means that the term was correctly identified by the MetaMap NLP process but it was discarded by our validation process. This problem was very typical in the case of acronyms within diagnostic tests. The sources used to generate the list of validation terms were impoverished for terms related to diagnostic tests (and diagnostic tests results) resulting in a high number of false negatives.

Another problem was detected with "classical" false negatives (a term that has been incorrectly rejected). Several terms were not identified by the NLP process. Most of these terms are "sentences" or composite phrases, which complicates their identification by the NLP process. Finally, there were very few false positives.

---

[8]https://github.com/alejandrorg/medlineplus2ddx.

[9]https://github.com/alejandrorg/medlineplus2ddx/blob/master/diseasesList.txt.

[10]https://github.com/alejandrorg/medlineplus2ddx/blob/master/Results.xlsx.

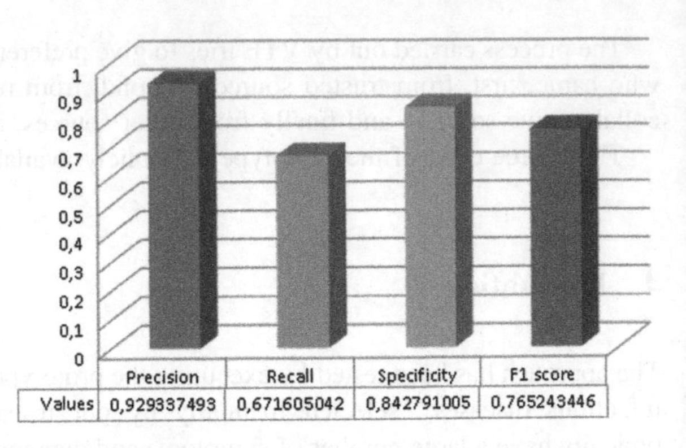


**Fig. 2** Statistical results.

## 5 Conclusions and Future Work

This paper presents a novel approach to extracting diagnostic clinical findings from MedlinePlus articles about infectious diseases. Evaluation of the prototype developed reveals that the proposed method is accurate enough to be used to extract diagnostically-relevant terms from several different sources of information (Web pages, clinical texts, etc.) by applying necessary modifications to the crawler. However, the evaluation also reveals that improvements could enhance performance.

As future work we consider that an expansion of the VTE procedure through adding new data sources to increase the number of validation terms might improve the quality of the results through reduction of false negatives. Another line would be to use a different NLP tool to process the input texts, for example Apache cTakes. Finally, we plan to extend our work to the domain of treatment information to enrich the knowledge extracted.

**Acknowledgments** Alejandro Rodríguez González's and Mark Wilkinson's work is supported by Isaac Peral Programme of the UPM. Marcos Martínez-Romero work has been supported by a Postdoc Fellowship from the Xunta de Galicia, Spain (ref. POS-A/2013/197).

## References

1. Tsumoto, S.: Automated extraction of medical expert system rules from clinical databases based on rough set theory. Inf. Sci. **12**(1–4), 67–84 (1998)
2. Tan, K.C., Yu, Q., Heng, C.M., Lee, T.H.: Evolutionary computing for knowledge discovery in medical diagnosis. Artif. Intell. Med. **27**, 129–154 (2003)
3. Hahn, U., Romacker, M., Schulz, S.: medSynDiKATe—a natural language system for the extraction of medical information from findings reports. Int. J. Med. Inf. **67**(1–3), 63–74 (2002)
4. Amaral, M.B., Roberts, A., Rector, A.L.: NLP techniques associated with the OpenGALEN ontology for semi-automatic textual extraction of medical knowledge: abstracting and mapping equivalent linguistic and logical constructs. In: Proceedings if the AMIA Annual Symposium, pp. 76–80 (2000)

5. Aronson, A.R.: Effective mapping of biomedical text to the UMLS Metathesaurus: the MetaMap program. In: Proceedings of the AMIA Annual Symposium, pp. 17–21 (2001)
6. Hamosh, A., Scott, A.F., Amberger, J.S., Bocchini, C.A., McKusick, V.A.: Online Mendelian Inheritance in Man (OMIM), a knowledgebase of human genes and genetic disorders. Nucleic Acids Res. **33**(1), 514–517 (2005)
7. Köhler, S., et al.: The human phenotype ontology project: linking molecular biology and disease through phenotype data. Nucleic Acids Res. **42**(D1), 966–974 (2014)
8. Bodenreider, O.: The unified medical language system (UMLS): integrating biomedical terminology. Nucleic Acids Res. **32**(1), 267–270 (2004)
9. Okumura, T., Aramaki, E., Tateisi, Y.: Clinical vocabulary and clinical finding concepts in medical literature. In: Proceedings of the International Joint Conference on Natural Language Processing Workshop on Natural Language Processing for Medical and Healthcare Fields, pp. 7–13 (2013)
10. Okumura, T., Tateisi, Y.: A lightweight approach for extracting disease-symptom relation with MetaMap toward automated generation of Disease Knowledge Base. Health Inf. Sci. 164–172 (2012)
11. Wu, Y., Denny, J.C., Rosenbloom, S.T., Miller, R.A., Giuse, D.A., Xu, H.A.: comparative study of current clinical natural language processing systems on handling abbreviations in discharge summaries. In: Proceedings of the AMIA Annual Symposium, pp. 997–1003 (2012)
12. Denecke, K.: Extracting medical Concepts from medical social media with clinical NLP tools: a qualitative study. In: Proceedings of the Fourth Workshop on Building and Evaluation Resources for Health and Biomedical Text Processing (2014)
13. Rodríguez-González, A., Martinez-Romero, M., Egaña-Aranguren, M., Wilkinson, M.D.: Nanopublishing clinical diagnoses: tracking diagnostic knowledge base content and utilization. In: IEEE 27th International Symposium on Computer-Based Medical Systems (CBMS), pp. 335–340 (2014)
14. Zhou, X.Z., Menche, J., Barabási, A.-L., Sharma, A.: Human symptoms–disease network. Nat. Commun. **5** (2013)

# A Text Mining Approach for the Extraction of Kinetic Information from Literature

**Ana Alão Freitas, Hugo Costa, Miguel Rocha and Isabel Rocha**

**Abstract** Systems biology has fostered interest in the use of kinetic models to better understand the dynamic behavior of metabolic networks in a wide variety of conditions. Unfortunately, in most cases, data available in different databases are not sufficient for the development of such models, since a significant part of the relevant information is still scattered in the literature. Thus, it becomes essential to develop specific and powerful text mining tools towards this aim. In this context, this work has as main objective the development of a text mining tool to extract, from scientific literature, kinetic parameters, their respective values and their relations with enzymes and metabolites. The pipeline proposed integrates the development of a novel plug-in over the text mining tool @Note2. Overall, the results validate the developed approach.

**Keywords** Enzyme kinetics · Metabolic models · Text mining · Name entity recognition · Relation extraction · Databases

## 1 Introduction

Several areas of science and industry are increasingly interested in the use of metabolic models, since they allow a variety of *in silico* simulations under different experimental conditions, and provide tools for strain optimization towards the increased production of compounds of interest, among other applications. In most cases, genome-scale stoichiometric metabolic models are used, allowing to perform accurate simulations under steady state conditions [6, 7]. However, to better understand the dynamic behavior of metabolic systems in a wide variety of conditions, it is imperative to develop dynamic kinetic models of cellular metabolism. These

A. A. Freitas · M. Rocha (✉) · I. Rocha
Centre Biological Engineering, School of Engineering University of Minho, Braga, Portugal
e-mail: irocha@deb.uminho.pt

H. Costa
SilicoLife Lda, Rua Do Canastreiro, 15, 4715-387 Braga, Portugal

 
R. Overbeek et al. (eds.), *9th International Conference on Practical Applications of Computational Biology and Bioinformatics*, Advances in Intelligent Systems and Computing 375, DOI 10.1007/978-3-319-19776-0_10

models describe the mass balances for each metabolite in the system [3, 7]. Despite the existing large number of databases (e.g. BRENDA [5], SABIO-RK [17], ExPASy [8], MetaCyc [2], etc.), available data on kinetic structures and parameters are not sufficient for the development of such models and a large amount of relevant information still resides in the biomedical literature [12].

In recent years, interest in biomedical text mining (BioTM) has increased, as a way to extract automatically meaningful knowledge from unstructured texts [1]. Typical BioTM tasks are divided in two main areas: Information Retrieval (IR) and Information Extraction (IE), which is further sub-divided in two major sub-areas: Named Entity Recognition (NER) and Relation/Events Extraction (RE) [4, 16] (Fig. 1). To summarize, a set of relevant publications can be retrieved, searching in bibliographic repositories and form a document set (corpus). This can be submitted to an NER process where biological entities are tagged with classes (e.g. genes, proteins, compounds). This annotated corpus can be further submitted to a RE process to identify relations between entities [13, 16].

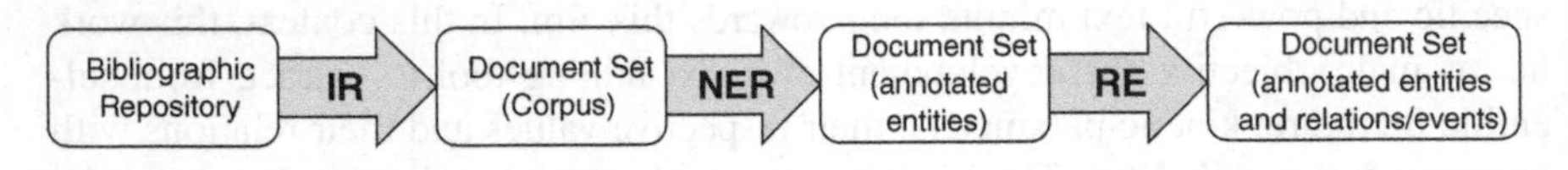


**Fig. 1** Generic pipeline for BioTM

There are currently very few BioTM tools specifically designed to extract kinetic data from the scientific literature. The BRENDA database developed a new resource, KENDA (Kinetic ENzyme DAta), which uses rule and dictionary-based approaches to extract kinetic values and expressions from literature. However, it only analyzes abstracts and titles [15]. Using a similar approach, the KID algorithm extracts kinetic information for enzymes from abstracts and automatically generates a database [10]. Other approaches based on machine learning [14] have also been focused on abstracts, therefore excluding the information contained in full text documents.

The main goal of this work is the development of a BioTM pipeline for the automatic identification and collection of kinetic data from literature, implemented as the KineticRE plug-in for the framework *@Note2* [11]. It works over annotated documents that result from an NER process. Based on those annotations, it looks for information on kinetic parameters, materialized by specific relations between the different annotated entities (kinetic parameters, values, enzymes and metabolites). The KineticRE algorithm is a rule-based RE process that returns a relation set ranked by score. The final objective is to obtain an ordered list of relations per document, which can be analyzed and may allow the user to select the parts of interest of the various documents, thereby decreasing the time necessary to collect relevant information.

## 2 @Note2

*@Note2* (http://www.anote-project.org/) is a multi-platform BioTM workbench, fully written in Java and that uses a MySQL database, which copes with the most important IR, NER and RE tasks. The framework is implemented over the AI-Bench framework, thus following the Model-View-Controller (MVC) paradigm [11]. *@Note2* is organized into three main functional modules (Fig. 2): Publication Manager Module (PMM), Corpora Module (CM) and Resources Module (RM).

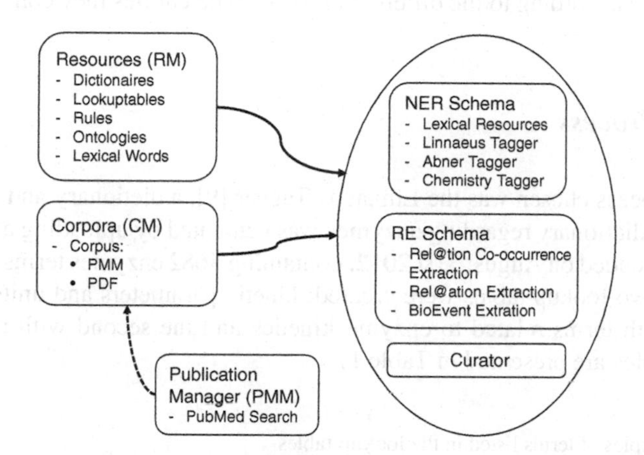


**Fig. 2** *@Note2* structure

The PMM is responsible for the IR step and has the functionality of searching and retrieving publications from repositories like PubMed, based on queries. The CM can receive those publications or import PDF files to form a Corpus. The RM is responsible for the management of lexical resources used in some IE processes, which include dictionaries, rules, ontologies, lookup tables and word lists. Different IE processes can be applied to a corpus, using some of the resources, to identify and extract bio-entities and their relationships. The Corpus processes sub-module contains three main processes: the NER process pool contains the Lexical resources and Linnaeus tagger [9], a hybrid dictionary and rule-based algorithm, the rule-based Chemistry tagger and the Machine Learning based ABNER tagger; the RE process pool contains a co-occurrence extraction process, a linguistic-based algorithm (Rel@tion) and a Machine learning approach; the Curator process allows to add/edit/remove entities and relations from IE results and to start a manual curation process [11].

## 3 Kinetic RE Pipeline

Since the RE processes previously available on *@Note2* are not adequate to extract kinetic information, we developed new specific RE algorithms, taking advantage on some of the *@Note* existing features. This new process works based on kinetic parameter values and names, enzymes and metabolites that are annotated using NER dictionaries and a rule-based process.

In the context of this work, a relation is defined as a set of entities from different classes, co-occurring in a sentence, which has a score. A basic relation occurs when an entity from the class values is found, followed by an entity from the class units. This basic relation, also named as a pair, is the basis to look for more complex relations. To form more complex relations, entities from the classes kinetic parameters, enzymes or metabolites have to be found before or after the previous pair. Relations receive a score according to the different classes of the entities they contain.

## *3.1 NER Process*

The NER process chosen was the Linnaeus Tagger [9], a dictionary and rule-based process. The dictionary regarding enzymes was generated by uploading a BRENDA [5] file downloaded on August 31st 2012, containing 4682 enzymes terms and 55743 synonyms. Two lookup tables were created: kinetic parameters and units. The first was filled with terms related to enzyme kinetics and the second with units terms (some examples are presented in Table 1).

**Table 1** Examples of terms listed in the lookup tables

| Units terms | Kinetic parameters terms |
|---|---|
| μmol; μmol-1; μmol -1 | KM; km; Km; kM |
| g mL -1; g mL-1; g mL - 1 | KI; ki; Ki; kI |
| g L -1; g l -1; g L-1; g l-1 | Vmax; vmax; V max; v max |
| Kcal / mol; Kcal/mol | Kcat/Km; Kcat / Km; kcat/km; kcat / km |
| (g dry weight) -1 h -1; (g dry weight)-1 h-1 | Inhibitor; Inhibition; inhibitor; inhibition |

Rules make use of Java regular expressions to identify specific patterns. From Table 2, it is possible to analyze some of the rules created.

**Table 2** Some regular expressions of the rules created and examples of instances

| Nr | Regular expressions (Java) | Examples |
|---|---|---|
| 1 | ?=\s+(V)\s+ | V |
| 2 | ?=\s+(-<br>0,1\d+0̇,1\d*\s0,1\s0,1+̣0̇,1\d*)\s+ | 7.8 23 ; -45 1.23 |
| 3 | \s(-0,1\d+0̇,1\d*\s0,1x\s0,110exp-0,1\d+) | 9.99x10exp9; 7x10exp-6 |
| 4 | ?=(\s+(-0,1\d+0̇\d+)\s+) | ]-99.99999; 99.99999[ |
| 5 | ?=(\s+(-0,1+̣)\s+) | ]-99999; 9999[ |

Lastly, an ontology was created, where the process is identical to a dictionary creation. In this case, a list of metabolites will be created from the ChEBI database. An obo file, with release 114 was downloaded (April 3rd 2014) containing 45436 terms and 266644 synonyms. To each term, an id, a name and synonyms are associated.

## *3.2 Identification of Relations*

The lexical resources used in the NER process need to be mapped by the user to the corresponding class (values, units, kinetic parameters, enzymes and metabolites), being the score values defined for each class. These will be used during the RE process to evaluate each relation identified.

The algorithm is depicted in Fig. 3. The search for a relation starts with the identification of values and units annotated together, where each occurrence found is considered a pair with minimum score assigned. For example, as shown in Fig. 4, two values (grey and underline) were annotated as belonging to the value class, but only the first is followed by an entity of Unit class (grey and italic), so only one pair will be identified.

**Input:** Annotated corpus (NER result), entities classes and corresponding scores.
**Algorithm:**

```
FOR each doc from the corpus:
    Split in sentences;
    FOR each sentence:
        Search for values annotated close to annotated units (distance between
        1 and 3 spaces);
            FOR each pair (value/unit):
                Consider a basic relation and assign the minimum score;
                Search for other entities annotated (metabolites, enzymes and
                kinetic parameters classes) on the left or right of the pair;
                Create a new relation (pair + new entities found);
                Calculate the relation score.
```

**Output:** List of complex relations with the corresponding score.

**Fig. 3** Algorithm for relation extraction in KineticRE

… activity was determined in crude extracts prepared in breaking buffer, 100 *mM* and pH 7.0.

PAIR
(value – unit)

**Fig. 4** Example of a sentence with only one value-unit pair

If only one pair is identified in the sentence, the whole sentence is considered the scope for the possible relations. If the algorithm does not find entities from the

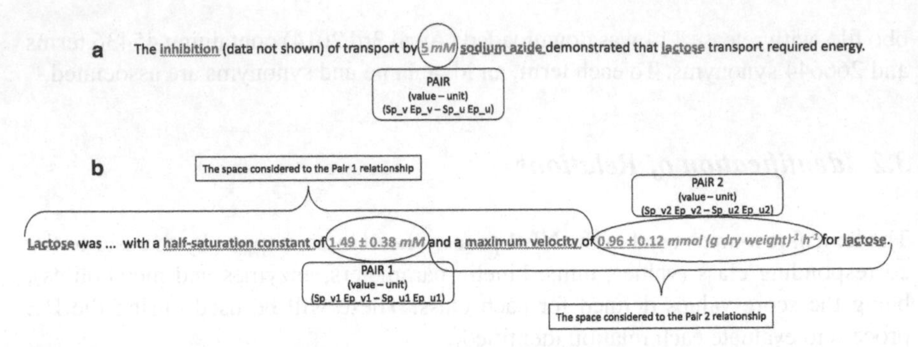


**Fig. 5** Example of sentences: with one value-unit pair (a) and with two value-unit pairs identified (b). Annotated entities from metabolites (dash long underline) and kinetic parameters (wavy underline) classes. SP_v and Ep_v are respectively start and end position of the value in the sentence; Sp_u and Ep_u represents the same for the unit parameter

other classes in the sentence, the pair is considered as the simplest relation and given the minimum score. Otherwise, when entities from the classes kinetic parameters, enzymes or metabolites are found either on the left (positions are lower than the pair/value start position Sp_v in Fig. 5.a), or on the right (positions are higher than pair/unit end position Ep_u) the score is increased. For each entity, the score to add is the one assigned to its class in the configuration of the algorithm.

Considering the following score values: pair→10; metabolites→100; enzymes→ 1000 and kinetic parameters→10000, the example in Fig. 5.a will receive a score of 10210 (10 + 2*100 + 10000). Clearly, for the computation of the scores, the greater the number of annotated entities of different classes, the higher will be the score assigned to the relation. If the sentence has two pairs (Fig. 5.b), for the first pair the scope to find a relation will be between the beginning of the sentence and the start position of the second pair (Sp_v2), while for the second it will be between the end position of the first pair (Ep_u1) and the end of the sentence. Using the scores defined above, the first relation gets a score of 20110 and the second a score of 10110.

It is possible to realize that the information annotated between pairs will be added to both relations, creating redundancy. In this case, it is obvious that the second parameter (wavy underline) belongs to the pair 2, but since there is no straight way to decide this kind of separation, it was decided to allocate the information in the middle to both pairs, with the intention of minimizing the probability of losing important information.

# 4 Implementation

The algorithm developed was integrated in *@Note2*, using functionalities already implemented, like the conversion from PDF to text and the overall NER process. As shown in Fig. 6, the algorithms receives an annotated corpus generated by an NER process, maps the lexical resources to the different classes and the result is a list of relations with scores, as well as a corpus with annotated entities and relations.

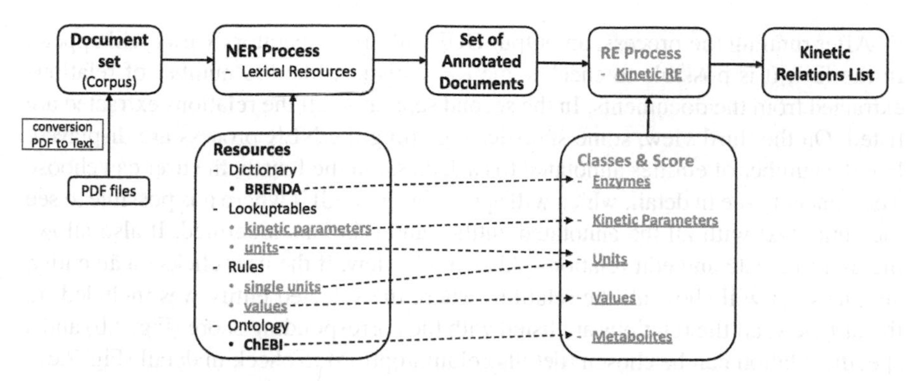


**Fig. 6** Main steps of the pipeline. Shown underlined are the new resources that were created in *@Note2* in the scope of this work

This plug-in can be selected within the RE options on the clipboard. Two input GUIs are then launched, in which the user has to choose the NER process and do the mapping between the resources used in NER and the classes available for the RE. For example, the resource unit (lookup table) and the resource unit_rule (set of rules) should be mapped to the RE class Unit, as shown in Fig. 7.a.

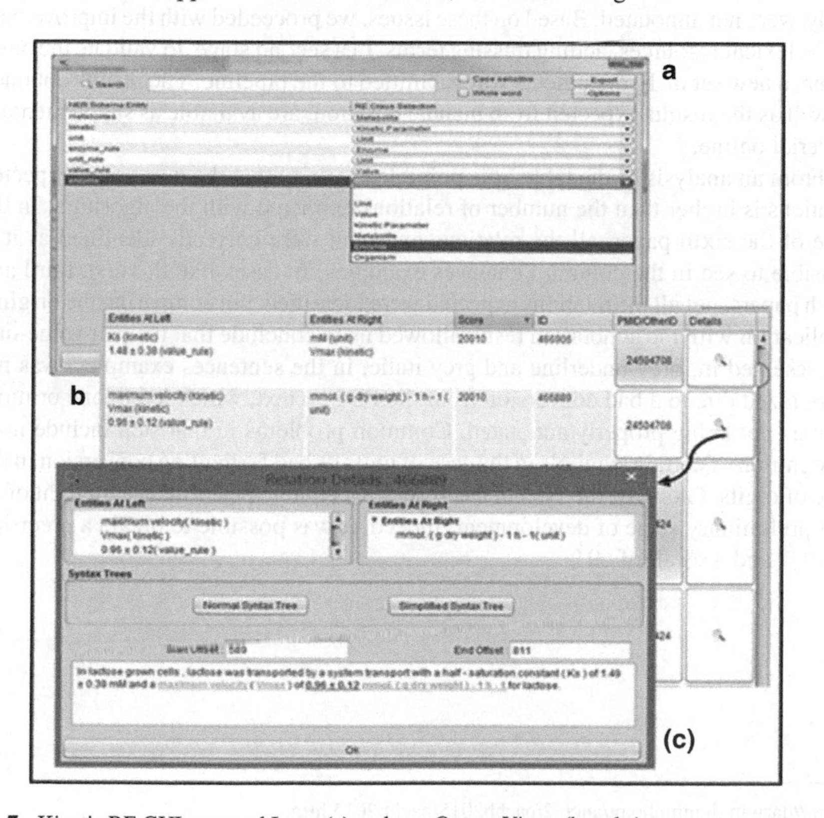


**Fig. 7** Kinetic RE GUIs: second Input (a) and two Output Views (b and c)

After running the process, an output GUI with five separators/views will appear. In the first, it is possible to check some RE statistics, like the number of relations extracted from the documents. In the second separator, all the relations extracted are listed. On the third view, some statistics regarding the NER process are displayed, like the number of entities annotated to each class. In the fourth, the user can choose a document to see in detail, which will appear in a new GUI where it is possible to see the entire text with all the annotated entities and relations identified. It also allows the user to create and edit relations. Also in this view, if the user clicks on an entity, another GUI will show all the relations where the selected entity was included. In the last view, all the relations are listed with the corresponding score (Fig. 7.b) and a specific relation can be chosen (details column/option) to check in detail (Fig. 7.c).

## 5 Results and Discussion

Initially, a set of 10 articles on Kluyveromyces lactis was selected to guide the creation of the different lexical resources and to validate the NER process. During the validation, it was detected that some terms of interest, that have been identified manually, were not annotated. Based on these issues, we proceeded with the improvement of the lexical resources, adding missing terms. In a second stage, to validate the algorithm, a new set of 10 documents was submitted to the pipeline. The results obtained as well as the results expected from manual curation, are available as supplementary material online.[1]

From an analysis of the table, it is possible to verify that the number of expected relations is higher than the number of relations extracted with the algorithm. In the case of the sixth paper, all the relations expected were correctly identified as it is possible to see in the column Sentences examples. In cases like the first, third and ninth papers, not all the relations expected were identified, but comparing the original publication with the annotation result allowed us to conclude that the pair value-unit (represented in, grey underline and grey italic, in the sentences examples) was not recognized due to a bad conversion from the PDF to text, which led to one or more entities not being properly annotated. Common problems in that step include non-recognition of certain symbols in the case of units or bad formatting conversion in the case of digits. Globally, the system has shown acceptable performance, in light of its still preliminary stage of development. Indeed, it was possible to obtain a precision of 79 % and a recall of 69 %.

[1]http://darwin.di.uminho.pt/anote2/pacbb2015/pacbb2015.htm.

## 6 Conclusions and Further Work

In this work, we have proposed and validated a pipeline to collect kinetic data from literature, working over the text mining tool *@Note2*, for which we developed a plug-in that is made available for the community. Overall, the preliminary results are quite promising. The PDF conversion to text is a core task, but also one of the more problematic. Despite the improvements done during this work, the results are far from perfect and leave room for further improvements. The same can be said for the resources used, which can be altered to improve the identification of entities. For example, this can be achieved adding terms to the units and kinetic parameters lookup tables, or fitting the rules for a better detection of values. The development and validation of this tool will continue along these lines, considering all other features that can improve the results.

**Acknowledgments** The work was funded by National Funds through the FCT (Portuguese Foundation for Science and Technology) within project ref. PTDC/QUI-BIQ/119657/2010 Finding the naturally evolved design principles of prevalent metabolic circuits. The authors would also like to thank the FCT Strategic Project PEst-OE/EQB/ LA0023/2013 and the Projects BioInd - Biotechnology and Bioengineering for improved Industrial and Agro-Food processes, REF. NORTE-07-0124-FEDER-000028 and PEM Metabolic Engineering Platform, project number 23060, both co-funded by the Programa Operacional Regional do Norte (ON.2 O Novo Norte), QREN, FEDER.

## References

1. Ananiadou, S., Kell, D.B., Tsujii, J.-I.: Text mining and its potential applications in systems biology. Trends Biotechnol. **24**(12), 9–571 (2006)
2. Caspi, R., Altman, T., Dreher, K., Fulcher, C.A., Subhraveti, P., Keseler, I.M., Kothari, A., Krummenacker, M., Latendresse, M., Mueller, L.A., Ong, Q., Paley, S., Pujar, A., Shearer, A.G., Travers, M., Weerasinghe, D., Zhang, P., Karp, P.D.: The MetaCyc database of metabolic pathways and enzymes and the BioCyc collection of pathway/genome databases. Nucleic Acids Res. **40**(Database issue):D742–D753 (2012)
3. Chassagnole, C., Noisommit-Rizzi, N., Schmid, J.W., Mauch, K., Reuss, M.: Dynamic modeling of the central carbon metabolism of Escherichia coli. Biotechnol. Bioeng. **79**(1), 53–73 (2002)
4. Cohen, K.B., Hunter, L.: Getting started in text mining. PLoS Comput. Biol. **4**(1), e20 (2008)
5. Dis, G. F., Schomburg, I., Hofmann, O., Baensch, C.: Enzyme data and metabolic information : BRENDA, a resource for research in biology, biochemistry, and medicine, pp. 3–4 (2000)
6. Durot, M., Bourguignon, P.-Y., Schachter, V.: Genome-scale models of bacterial metabolism: reconstruction and applications. FEMS Microbiol. Rev. **33**(1), 90–164 (2009)
7. Edwards, J.S., Palsson, B.O.: Robustness analysis of the Escherichia coli metabolic network. Biotechnol. Prog. **16**(6), 927–939 (2000)
8. Gasteiger, E.: ExPASy: the proteomics server for in-depth protein knowledge and analysis. Nucleic Acids Res. **31**(13), 3784–3788 (2003)
9. Gerner, M., Nenadic, G., Bergman, C.M.: LINNAEUS : a species name identification system for biomedical literature (2010)
10. Heinen, S., Thielen, B., Schomburg, D.: KID-an algorithm for fast and efficient text mining used to automatically generate a database containing kinetic information of enzymes. BMC Bioinf. **11**, 375 (2010)

11. Lourenço, A., Carreira, R., Carneiro, S., Maia, P., Glez-Peña, D., Fdez-Riverola, F., Ferreira, E.C., Rocha, I., Rocha, M.: @Note: a workbench for biomedical text mining. J. Biomed. Inf. **42**(4), 20–710 (2009)
12. Patil, K.R., Åkesson, M., Nielsen, J.: Use of genome-scale microbial models for metabolic engineering. Curr. Opin. Biotechnol. **15**(1), 64–69 (2004)
13. Rodriguez-Esteban, R.: Biomedical text mining and its applications. PLoS Comput. Biol. **5**(12), e1000597 (2009)
14. Schmeier, S., Kowald, A., Klipp, E., Leser, U.L.F.: Finding kinetic parameters using text mining. **8**(2), 131–153 (2004)
15. Schomburg, I., Chang, A., Placzek, S., Söhngen, C., Rother, M., Lang, M., Munaretto, C., Ulas, S., Stelzer, M., Grote, A., Scheer, M., Schomburg, D.: BRENDA in 2013: integrated reactions, kinetic data, enzyme function data, improved disease classification: new options and contents in BRENDA. Nucleic Acids Res. **41**(Database issue):D764–D772 (2013)
16. Shatkay, H., Craven, M.: Mining the biomedical literature. MIT Press (2012)
17. Wittig, U., Golebiewski, M., Kania, R., Krebs, O., Mir, S., Weidemann, A., Anstein, S., Saric, J., Rojas, I.: SABIO-RK : integration and curation of reaction kinetics data, pp. 94–103 (2006)

# A Novel Search Engine Supporting Specific Drug Queries and Literature Management

**Alberto G. Jácome, Florentino Fdez-Riverola and Anália Lourenço**

**Abstract** The growing concern for acquired microbial resistance is promoting the publication of a large number of clinical and biological antimicrobial studies. Most of these publications can be obtained by searching the PubMed database, but the broad scope and huge size of this collection make the search challenging, and time consuming. This paper presents an advanced search engine for the screening of up-to-date information on drug-related experimental studies. The main contributions lay on the resource-oriented architecture and the semantic analysis of the documents. The RESTful API enables the use of the searchable collection by different user interfaces whereas text mining tools support domain-specific document labeling, scoring and indexing. A small search engine demo indexing articles on antimicrobial peptide research is available at http://sing.ei.uvigo.es/sds/. The source code is also accessible from the same homepage and freely available under MIT License.

**Keywords** Drugs · Experimental testing · Vertical search engine · Text mining · Document annotation · Web application

A.G. Jácome · F. Fdez-Riverola · A. Lourenço (✉)
ESEI - Escuela Superior de Ingeniería Informática Edificio Politécnico, Campus Universitario as Lagoas S/N, Universidad de Vigo, 32004 Ourense, Spain
e-mail: analia@uvigo.es; analia@ceb.uminho.pt

A.G. Jácome
e-mail: agjacome@esei.uvigo.es

F. Fdez-Riverola
e-mail: riverola@uvigo.es

A. Lourenço
Centre of Biological Engineering, University of Minho, Campus de Gualtar, 4710-057 Braga, Portugal



R. Overbeek et al. (eds.), *9th International Conference on Practical Applications of Computational Biology and Bioinformatics*, Advances in Intelligent Systems and Computing 375, DOI 10.1007/978-3-319-19776-0_11

## 1 Introduction

Infectious diseases retain a prominent position as major worldwide cause of morbidity and mortality in a wide range of patients, and incidence of drug resistance in clinical isolates is growing [1, 2]. Increased knowledge about the molecular mechanisms of virulence is important to develop our understanding of pathogenesis and establish innovative treatments for infection. In this context, the investigation of antimicrobial agents with alternative mechanisms of action and the evaluation of the combinatory effects of antimicrobial agents against multidrug-resistant strains are pressing biomedical research goals [3].

Drug-related databases such as DrugBank [4] and CAMP [5], databases on protein-compound interactions such as STITCH [6], and web resources such as the list of minimum inhibitory concentrations of the European Committee on Antimicrobial Susceptibility Testing (http://mic.eucast.org/Eucast2/), can help researchers in planning new studies and understanding the results in hand. However, most information lays on scientific literature and requires additional search and curation abilities. For example, literature searches are needed to find out experimental and trial results on new drugs and drug combinations, and pharmacokinetics details.

PubMed is the resource of reference to search for biomedical bibliography, but its broad scope and the increasing volume of biomedical literature pose a challenge to the satisfaction of individual user needs [7]. The list of results for a PubMed query can easily contain hundreds or thousands of documents, and reading through all the returned titles and abstracts for useful information is inefficient, especially as a routine. Users need to be sensitive to the operation of Boolean-like queries and identify the best keywords for their queries, in order to avoid the retrieval of a large number of uninteresting documents. Typically, this search experience is iterative and has to be adjusted to the continuous growth of the engine, i.e. more documents imply more indexed contents, and often, new terminology-domain associations to recognize and handle.

The development of advanced or domain-specific search engines is an attractive alternative to create customized and enhanced search experiences [8]. In particular, the use of semantic analysis or text mining approaches as an adjunct to, or in replacement of, Boolean searching may be useful. Drug-related portals exist, but they are mainly focused on reporting main drug information, or disease-specific implications, rather than supporting research topics and the navigation of experimental findings [9].

Within this context, the present paper introduces SDS (Smart Drug Search), a new search engine in assistance of antimicrobial research, in particular the screening of literature related to antimicrobial resistance, microbial virulence and topics alike. The innovation and main contribution of this engine is its flexible and highly extensible resource-oriented architecture, encompassing a resource abstraction layer, a resource management and control layer, and an open service interface layer. The search engine combines state-of-the-art web technologies and frameworks with text mining tools so that literature monitoring is active and efficient, and user experience may be easily customized and refined. That is, the topics covered by the search engine may be extended, new documental sources may be included, and document annotation and semantic navigation may be customized according to the audience being targeted. Moreover, a public RESTful application programming interface (API) provides access to the searchable document collection such that it can be easily accessed from different user interfaces. Overall, these capabilities make SDS an easy-to-implement tool that research groups and centers, and other bodies, may customize to make part of their Web services.

The next sections detail this architecture, highlighting the potential of semantic analysis in fine tuning document indexes and assessing document relevance, and describing the ability of the engine to adapt to common domain-specific user search behaviors.

## 2 Search Engine Architecture

Different strategies may be combined to improve the performance of a search engine when applied to specialized search tasks, namely semantic analysis and text mining. Previous works on the retrieval of drug-drug interaction and pharmacokinetics information have shown that text mining tools are able to recognize drug-related entities effectively [10]. The SDS engine applied some of these tools to annotate information about diseases, pathogens, drugs and mechanisms of action, and thus be able to prioritize and filter the document in the knowledge base. Drug-related searches may then be conducted based on the meaning of annotated terms rather than simply on their presence (allowing us to normalize synonyms and equivalent representations of a concept), and records may be scored according to a given probability of relevance.

In terms of software development, SDS was implemented as a web-based application, being designed as a common three-layered architecture where the bottom layer is responsible for accessing the database, the middle layer supports the main functionalities of the web search engine and the top layer is in charge of providing the end-user interface (see Fig. 1). Complementarily, SDS follows a flexible client-server model where each layer can be easily moved to a different machine accessible through the network, but with the web-based interface being always executed in the client side with the objective of relieving the server of rendering user views.

The SDS system was developed using the Scala programming language (http://www.scala-lang.org/) in the server application (middle logic layer) and JavaScript in the client application (top presentation layer). The development of the Web interface was facilitated through the use of the model-view-controller framework AngularJS (https://angularjs.org/) together with the HTML and CSS framework Bootstrap (http://getbootstrap.com/). Complementarily, in the server-side application we used Play (https://www.playframework.com/) as a container framework,

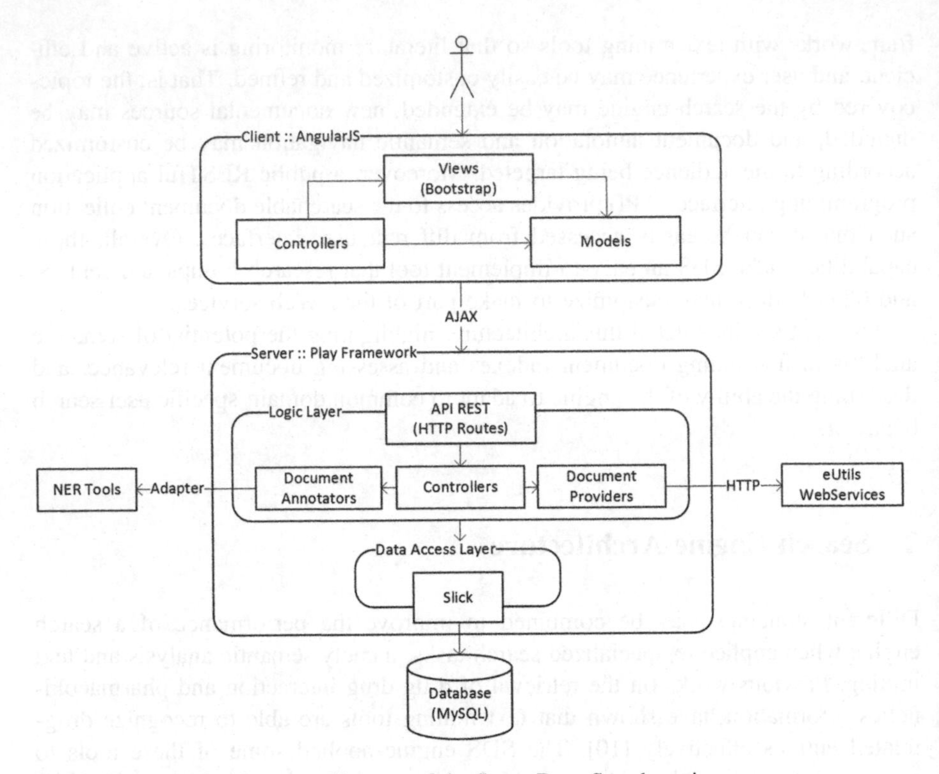


**Fig. 1** The general software architecture of the Smart Drug Search engine

Akka (http://akka.io/) to simplify the distribution and parallelization of different parts of the system, and Slick (http://slick.typesafe.com/) to facilitate the interaction with the relational database.

On top of the middle layer, SDS supports a RESTful API enabling a simple way to execute each operation implemented by the search engine. This can be directly done through a set of HTTP commands that make use of JSON as the data interchange format. While some of these requests are non-blocking, allowing the client to continue its execution without having to wait for a long-running operation, others block the client because of their intrinsic nature. A small subset of implemented functionalities is shown in Table 1.

Inside the logic layer resides the document provider and annotator subsystems (central part of Fig. 1), where all the current document analysis techniques are implemented. These subsystems use external resources to correctly operate, namely: the repositories where original documents are downloaded from and natural language processing (NLP) techniques to annotate the terms of interest of each document. Both modules have been consciously designed and developed to allow full extensibility. This means that new document repositories and novel NER (Named Entity Recognition) tools can be easily integrated into the search engine, and it is possible to activate and deactivate their operation through system

**Table 1** Main functionalities provided through the RESTful API of SDS

| Function | Description |
|---|---|
| *Search documents* | Searches documents that conform to some given terms.<br>`GET /api/search?query = terms&page = 1&count=10` |
| *Access single Document* | Retrieves all the data related to a single document.<br>`GET /api/document/1` |
| *List all Documents* | Lists all the documents stored at the system.<br>`GET /api/document` |
| *Annotate documents* | Annotates a given document. Requires administrator privileges and previous API login.<br>`POST /api/annotator/annotate`<br>Content: `{ ids: [1, 2, 3, 4] }` |
| *Download documents* | Downloads a set of documents from PubMed. Requires administrator privileges and previous API login.<br>`POST /api/provider/pubmed/download`<br>Content: `{ ids: [25542381, 25453271] }` |

configuration. All these sub-modules run while SDS is operative, allowing the retrieval and annotation of new articles and the update of the database index without affecting end usage.

In the current version of the search engine, all the documents are retrieved from PubMed through the *Entrez Programming Utilities* (EUtils), belonging to the National Centre for Biotechnology Information. In particular, the EUtils SOAP API allows direct access from code by just generating the required functionality through the publicly available WSDL and XSD files. At the moment, SDS is configured to download only the abstract portion of each requested article, since republishing full-text articles is likely to infringe journal copyrights. Those abstracts are persisted in a MySQL database that will serve as a search index once they are annotated.

Regarding document annotation, the module implemented by SDS is composed of different NER tools able to detect and normalize to a common representation terms relevant to the drugs domain. At the moment, there are three tools integrated into SDS: ABNER (*A Biomedical Named Entity Recognizer*) [11], OSCAR (*Open Source Chemistry Analysis Routines*) [12] and Linnaeus [13]. All of them can be used as Java libraries and be accessed straightforwardly from within the Scala code.

On one hand, ABNER allows the system to analyze terms related to molecular biology (e.g., proteins, DNA, cell types, genes, etc.), but it does not handle their normalization. This fact causes that two synonyms will not be merged into a unique common representation, and they may not be used interchangeably in user queries. So, it is likely that ABNER will be replaced by an alternative tool in future versions of SDS. On the other hand, both Linnaeus and OSCAR perform normalization of synonym terms, which make them a good fit into the search engine. Linnaeus provides the NCBI Taxonomy identifier of each annotated species as a form of normalization, so in order to completely normalize it in the system, the scientific name provided by NCBI Taxonomy is used, OSCAR is used to recognize chemical compound names and chemical formulae and normalize them to a standard notation. OSCAR supports different notation formats, the one used in SDS is the InChI (*IUPAC International Chemical Identifier*).

While our current pursue with this architecture is to create a drug-based search engine, its fully extensible design permits any interested developer to reuse it with different annotation and document provider modules. This implies that new search engines, related or not to drug queries, can be created with ease by just including new NER tools and new remote database parsers. Current medical search engines do not allow this and are designed in order to support only one kind of searches, making necessary to design and implement a complete new search engine if a new kind is to be supported.

## 3 Usage Example

End-users of the SDS engine can easily access the web interface. The public home page of the application presents an intuitive input box, where the main operation (searching) can be performed by just inserting a set of search terms (see Fig. 2). Those terms will be processed by the system in order to obtain the best matches within all the previously analyzed documents. When the engine finishes processing the user request, a list containing all the documents that correspond to the given search will be shown enriched with a set of normalized terms. Those terms are displayed with a color code that reflects the category in which they have been detected (e.g., species, protein, cell type, etc.), being also displayed as a tooltip in each term.

A document can be accessed through a previous search or by using the corresponding URL. Either way, the user interface will display the text of the document (only abstracts if they have been downloaded from PubMed) with all its relevant terms highlighted with the same color corresponding to the appropriate category. All the terms highlighted by SDS also present a tooltip where more information about them is given (i.e., its associated category and the normalized version of each term) as processed by the document analysis subsystem.

Figure 2 shows an example of a simple search and some administration details. The top left screenshot shows a result list of documents obtained with the query 'antimicrobial peptide', where a list of matching terms can be seen along each document title. Bottom left image continues this process by accessing one of those previous documents, where the full abstract and a link to PubMed are provided. All the relevant terms in the abstract are highlighted and tooltips provide further details.

On the administration area, new documents can be retrieved from PubMed, submitted to SDS annotation module and incorporated to the database of the engine. In Fig. 2, the top right figure shows the panel that coordinates the retrieval of PubMed documents, and the bottom right figure shows a list of documents, where some of them are already annotated by the system, some are not annotated and others are being annotated at the time of the screenshot.

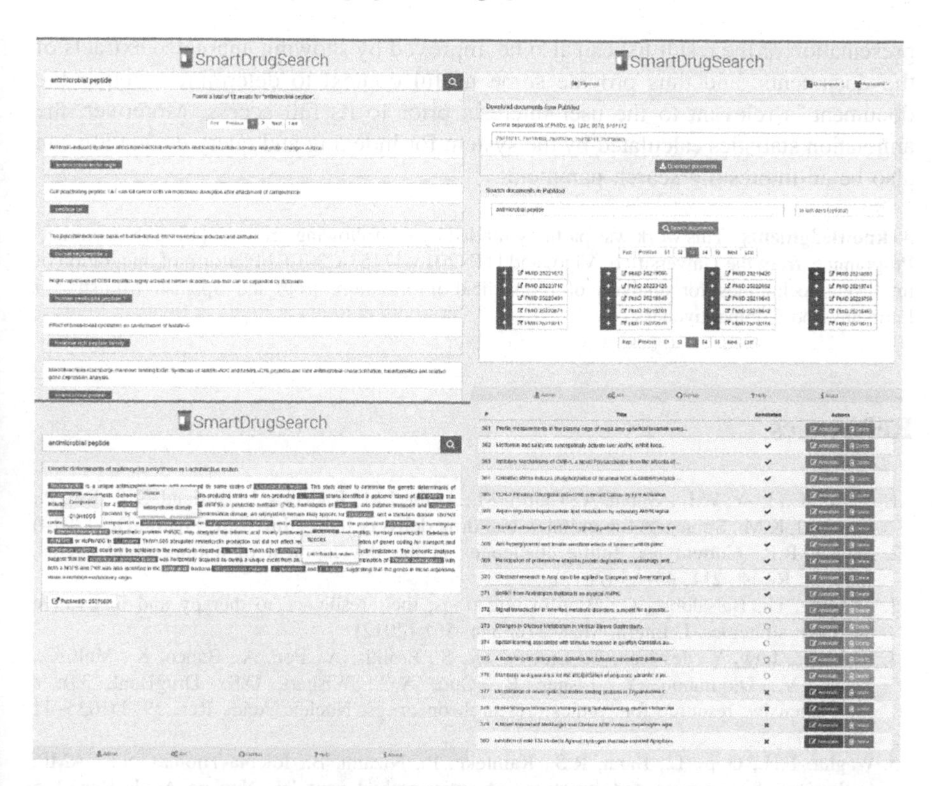


**Fig. 2** The two functionalities offered by the web interface of SDS: user query (on the left side) and search engine administration (on the right side)

## 4 Conclusions and Further Work

The drug-related search engine presented in this work is a free MIT licensed web-based application that helps researchers in the design of new drugs by providing advanced searching capabilities which aim to discover relevant information in scientific literature. The ability to auto-annotate, through text mining mechanisms, all the papers stored in the system is of great help to create a powerful search index that can normalize synonyms and different notations of the same term, also helping specialists to use whichever form they are more comfortable with. Moreover, the RESTful API provided by SDS helps to integrate all the features present in this application into already existing, or future, tools making this search engine a fully reusable software component.

This is, of course, a non-definitive version of the tool. There is a lot of future work, including enhancements and new features, being planned for following versions. Upcoming versions of the search engine will improve end-user experience by enabling advanced search parameters (e.g., to filter matching terms to a given category or set of categories, by publication date, etc.). Additionally, the presentation of the result list can also be improved by showing annotated extracts of the documents, and thus provide some useful context to determine whether the document is relevant to the user interests prior to its full access. Moreover, the annotation statistics calculated by the system for indexing and query evaluation can also be an interesting search parameter.

**Acknowledgments** This work was partially funded by the following projects: [14VI05] Contract-Programme, from the University of Vigo, and [TIN2013-47153-C3-3-R] Platform of integration of intelligent techniques for analysis of biomedical information, from the Spanish Ministry of Economy and Competitiveness.

## References

1. Martin, C., Low, W.L., Gupta, A., Cairul, M., Amin, I.M., Radecka, I., Britland, S.T., Raj, P., Kenward, K.M.: Strategies for antimicrobial drug delivery to biofilm. Curr. Pharm, Des (2014)
2. Rennie, R.P.: Current and future challenges in the development of antimicrobial agents. Antibiot. Resist. **211**, 45–65 (2012)
3. Römling, U., Balsalobre, C.: Biofilm infections, their resilience to therapy and innovative treatment strategies. J. Intern. Med. **272**, 541–561 (2012)
4. Knox, C., Law, V., Jewison, T., Liu, P., Ly, S., Frolkis, A., Pon, A., Banco, K., Mak, C., Neveu, V., Djoumbou, Y., Eisner, R., Guo, A.C., Wishart, D.S.: DrugBank 3.0: a comprehensive resource for "omics" research on drugs. Nucleic Acids Res. **39**, D1035–41 (2011)
5. Waghu, F.H., Gopi, L., Barai, R.S., Ramteke, P., Nizami, B., Idicula-Thomas, S.: CAMP: Collection of sequences and structures of antimicrobial peptides. Nucleic Acids Res. 1–5 (2013)
6. Kuhn, M., Szklarczyk, D., Franceschini, A., von Mering, C., Jensen, L.J., Bork, P.: STITCH 3: zooming in on protein-chemical interactions. Nucleic Acids Res. **40**, D876–D880 (2012)
7. Lu, Z.: PubMed and beyond: A survey of web tools for searching biomedical literature. Database **2011**, 1–13 (2011)
8. Pasche, E., Gobeill, J., Kreim, O., Oezdemir-Zaech, F., Vachon, T., Lovis, C., Ruch, P.: Development and tuning of an original search engine for patent libraries in medicinal chemistry. BMC Bioinform. **15**(Suppl 1), S15 (2014)
9. Galperin, M.Y., Rigden, D.J., Fernández-Suárez, X.M.: The 2015 Nucleic Acids Research Database Issue and molecular biology database collection. Nucleic Acids Res. **43**(Database issue):D1–D5 (2015). doi:10.1093/nar/gku1241
10. Percha, B., Altman, R.B.: Informatics confronts drug-drug interactions. Trends Pharmacol. Sci. **34**, 178–184 (2013)
11. Settles, B.: ABNER: an open source tool for automatically tagging genes, proteins, and other entity names in text. Bioinformatics **21**(14), 3191–3192 (2005)
12. Jessop, D.M., Adams, S.E., Willighagen, E.L., Hawizy, L., Murray-Rust, P.: OSCAR4: a flexible architecture for chemical text-mining. J. Cheminform. **3**(1), 41 (2011). doi:10.1186/1758-2946-3-41
13. Gerner, M., Nenadic, G., Bergman, C.M.: LINNAEUS: a species name identification system for biomedical literature. BMC Bioinform. **11**, 85 (2010). doi:10.1186/1471-2105-11-85

# Identification of a Putative Ganoderic Acid Pathway Enzyme in a *Ganoderma Australe* Transcriptome by Means of a Hidden Markov Model

**Germán López-Gartner, Daniel Agudelo-Valencia, Sergio Castaño, Gustavo A. Isaza, Luis F. Castillo, Mariana Sánchez and Jeferson Arango**

**Abstract** *Ganoderma australe* is a fungus widely used as a traditional medicine mainly in Eastern countries, but not studied *in silico* at the genomic level. This species is probably related to other well characterized fungus with similar properties, which may facilitate gene finding through comparative molecular analysis using appropriated bioinformatics tools. This paper aims to present a preliminary analysis of a *G. australe* transcriptome through some computational biology techniques implementing Hidden Markov Models (HMM) in order to predict a key putative enzyme (lanosterol synthase, EC 5.4.99.7) involved in the metabolic pathway of triterpenoids of therapeutic interest. The findings suggest that the HMM approach results more efficient than traditional comparisons by homology based on

G. López-Gartner (✉) · D. Agudelo-Valencia · M. Sánchez
Biology Sciences Department, GITIR Research Group,
Universidad de Caldas, Street 65 # 26–10, Manizales, Colombia
e-mail: german.lopez@ucaldas.edu.co

D. Agudelo-Valencia
e-mail: daniel.ag.valencia@gmail.com

M. Sánchez
e-mail: mari_salob@hotmail.com

S. Castaño · G.A. Isaza · L.F. Castillo · J. Arango
Systems and Informatics Department, GITIR Research Group,
Universidad de Caldas, Street 65 # 26–10, Manizales, Colombia
e-mail: andres3275@gmail.com

G.A. Isaza
e-mail: gustavo.isaza@ucaldas.edu.co

L.F. Castillo
e-mail: luis.castillo@ucaldas.edu.co

J. Arango
e-mail: jeferson.arango@ucaldas.edu.co



R. Overbeek et al. (eds.), *9th International Conference on Practical Applications of Computational Biology and Bioinformatics*, Advances in Intelligent Systems and Computing 375, DOI 10.1007/978-3-319-19776-0_12

methods of multiple sequences alignment. Here we report the first evidence of a putative lanosterol synthase protein being expressed in cell cultures of *G. australe*.

**Keywords** HMM bioinformatics • Ganoderma • Gene finding

## 1 Introduction

Bioinformatics is a discipline relatively new that helps with the discovery of biological information through the implementation of computational techniques. Very frequently, the tools of computational biology are based on well established mathematical o statistical models that make them more reliable for defining biological functions related to molecular structures of nucleic acids or protein sequences. Frequently the objective is to identify new products, e.g. molecules, compounds, metabolites, genes or proteins that represent some biotechnological interest.
In the search of new chemical compounds of biotechnological importance it is strategic to consider and explore all kinds of organisms in nature. In this endeavor it has been common practice in the genomic era to sequence, the genome, the transcriptome or the proteome of the species of interest. Once it has been done, an important goal of genome annotation projects is to identify all protein-coding genes. Currently, this task is too expensive and time consuming to be approached by experimental means in a standard molecular biology laboratory. Therefore, it appears much more feasible the implementation of gene finding by computational tools with algorithms using both intrinsic (ab initio, statistical) and extrinsic (sequence similarity) measures [1].

Increasingly, in the bioinformatics discipline protein function prediction becomes a central problem, gaining importance recently due to the rapid accumulation of biological sequencing data awaiting interpretation, especially biological function assignment to protein sequences is commonly required. Thus sequence data represents the bulk of this new wave and is the obvious target for consideration as input, as newly sequenced organisms often lack any other type of biological characterization.

Many methods for predicting biological function from molecular structure have been published. To this respect, typically the sequence-based methods used for gene finding and protein function assignment are traditional approaches such as identifying domains or finding Basic Local Alignment Search Tool (BLAST) or FASTA runs, which merely perform direct sequence to sequence comparisons between the query and the data bases [2]. These have been largely superseded by more powerful and sensitive profile/pattern-based methods, e.g. Hidden Markov Models (HMMs). These can provide important clues to function, even before the structure is obtained by crystallography technics in a wet laboratory stage.

Because of this, HMMs has become a remarkable formal foundation for making probabilistic models of linear sequence labeling problems. They are implemented by a diverse range of programs of common use in bioinformatics, including gene finding, profile searches, multiple sequence alignment and regulatory site identification. The most popular use of the HMM in molecular biology is as a probabilistic profile of a protein family, which is called a profile HMM. From a family of proteins (DNA, cDNA or RNA) a profile HMM can be made for searching a database for other members of the family. The profiles are constructed from the sequences of whole protein families, where the family can be defined in terms of three-dimensional structure, as in SUPERFAMILY, or in terms of function, as in Pfam. The primary advantage of the profile methods is that they provide greater sensitivity compared to simple sequence to sequence comparison because the profiles implicitly contain information on both which residues within the family are well conserved and which are the most variable [3].

In this study it is presented a general strategy of gene finding using a hidden Markov model (HMM), consisting in a statistical Markov model in which the system being modeled is assumed to be a Markov process with unobserved (hidden) states [4]. This approach was applied to a *"non model"* organism of biotechnological importance, the macromycetes fungus *Ganoderma australe*. The kingdom Fungi is considered one of the greatest resources in biodiversity given that they are present in almost all ecological niches, due to their important roles in ecosystems and their significant variety of contents in metabolites with different properties and a broad spectrum of chemical structures. It for these reasons fungi are amenable for the discovery of new applications in multiple fields of biotechnology [5, 6]. In particular, mushrooms belonging to the genus *Ganoderma* have been used as a traditional medicine in China and several Eastern countries for millennia, where it is considered "spiritual nature" or "mushroom of immortality" because of its nutraceutical and therapeutic properties [7]. Until recently, the main uses of *Ganoderma* fungi was restricted to folk medicine and the manufacture of dietary supplements that provide beneficial effects on health; however, relatively little is known about the mushroom's genomes, proteomes and secondary metabolism [8]. The uses of these fungi are very broad, going from industry bleaching paper pulp to the pharmaceutical industry and food production with nutraceutical properties [9].

Several species of *Ganoderma* have proved effective in industrial, biotechnological and therapeutic applications. For example, [10] reported the ability of *G. lucidum* for biodegradation of cyanide, in *G. applanatum* was investigated applying the bleaching and wastewater defenolization [11]. In [12] is studied the use of *G. australe* in the bioremediation of soils contaminated with recalcitrant organic compounds like lindano. Recently [13] studied the transcriptome of *G. lucidum* in order to identify the genes responsible for tolerance of this species to the presence of high concentrations of heavy metals such as cadmium. In therapeutic industry it has shown that some extracts and pure biomolecules of *G. lucidum* inhibit the growth of cancer cells and promote immune response in vivo [14] inducing the production of immunoglobulins. As newer biotechnological applications, *Ganoderma* is being employed for microsynthesis of nanoparticles for targeted modulation of antibiotic and antimicrobial activities [15] and anti-tumor effects [16].

In a previous work we obtained the complete sequence of the transcriptome of cultured cells (mycelium) from *G. australe* with the aim of a preliminary analysis of the transcriptome and finding genes of medical importance [17]. From this work we identified a candidate cDNA sequence belonging to the fungal immunomodulatory protein (FIP) family, which has shown promising results, tested and documented exhaustively for cancer therapy and other diseases that compromise the immune system [18, 19]

The transcriptome studied here consists of 712683 reads, 10,257 contigs (putative exons), 9.109 Isotigs (consensus sequences encoding putative proteins) and 8.170 isogroups (potential genes) [17]. Based on these data it was proceeded with the identification of new sequences of biomedical interest, including enzymes compromised in the ganoderic acids pathway. The ganoderic acids are a class of closely related triterpenoids (derivatives from lanosterol) found in *Ganoderma* mushrooms and responsible for a variety of putative pharmacological effects, including hepatoprotection and anti-tumor effects, proving efficiency in the control of prostate cancer and other tumorigenic cells growth through several immuno-modulatory induced effects, including interleukin enhancement mechanisms [20–22].

In spite of the promising biological effects of triterpenes derived metabolites from fungi like *Ganoderma*, with impact demonstrated especially in human health and medicine, important amounts of this compounds, bigger than those naturally produced, are needed in order to achieve a significant biological effect. Unfortunately, only very low amounts of triterpenes can be isolated from the mycelia and fruiting body of these fungi. Thus improvement of the qualitative and quantitative values of *Ganoderma* mushrooms has become an important issue as for a stable supply of a natural remedy for very critical health disorders [23]. One feasible solution to this problem is overproduction of triterpenes in cells genetically modify by means of genetic engineering and transgenics technology. As a first step to accomplish this, a better understanding of the roles and molecular characteristics of related components (genes, proteins, etc.) involved in triterpenes metabolic biosynthesis is necessary.

The main goal of this research is to identify in the obtained *G. australe* transcriptome putative sequences responsible for the expression of lanosterol synthase (EC 5.4.99.7), one of the key enzymes involved in the ganoderic acid synthesis, closely related to triterpenoids and directly derivative from lanosterol, using a bioinformatics strategy based on the HMM technique.

## 2 Materials and Methods

The first step in developing the ab initio gene finding algorithm was to build an appropriated multiple alignment model using a reasonable amount of evolutionary related sequences, let's say 10 or more homologous sequences with respect to that of interest, in this case, the enzyme lanosterol synthase. The second step was to build a statistical protein model or profile, i.e. Markov model, integrating the model into a pattern recognition algorithm and then it was used to scan all the sequences in the translated transcriptome.

We use a previously developed semantic and distributed service oriented-architecture for bioinformatics pipeline (*GITIRBio*), with special emphasis in the assembling, gene finding and annotation processes.[1] A general description of the pipeline is shown in the Fig. 1. The procedures implemented on this bioinformatics tool were based on the pipeline described in [24]. The prediction of genes is one of the most important processes of computational biology in the analysis of an organism after being sequenced, to achieve that, there are several bioinformatics tools that use different methods and data bases to identify DNA or amino acid sequences.

The score system is the core of the heuristic approach; it has 7 variables for DNA and 6 for amino acid sequences. For DNA the variables are match, mismatch, gap opening, gap deletion, gap insertion, transversion and transition and for amino acid are match, mismatch, gap opening, gap deletion, gap insertion, and a substitution matrix (PAM or blosum).

# 3 Results

## 3.1 Hidden Markov Model (HMM)

A Hidden Markov Model (HMM) was used to identify in the transcriptome of *G. australe* the possible sequences candidates to be involved in the synthesis of the ganoderic acids metabolism. In general terms, in a HMM applied to a biological context the codons are the observations, the amino acid sequence is the Markov chain and the switches from one genomic region to another are the state transitions: 5'UTR to CDS, CDS to intron, intron to CDS and CDS to 3'UTR.

The translated transcriptome was then compared against the probabilistic profile by *HMMER*[2] through the tool *hmmscan*, as shown in Fig. 2.

The Viterbi algorithm is a dynamic programming algorithm that finds the most probable path among several paths through the hidden states that lead to the given amino acid sequence. The time complexity of the Viterbi algorithm is $O((N^2)T)$, where $N$ are the number of states of the hidden markov model and $T$ the length of the sequence.

In the gene finding process we use the HMM method starting with the *G. australe* transcriptome. First the cDNA triplets were translated to amino acids. For that

[1]Castillo, L.F; López-Gartner, G.; Isaza, G.; Sánchez, M.; Arango, J.; Agudelo-Valencia, D.; Castaño, S. *GITIRBio*: A Semantic and Distributed Service Oriented-Architecture for Bioinformatics Pipeline. Sent to: Journal of Integrative Bioinformatics (JIB). In Review.

[2]HMMER web server: interactive sequence similarity searching R.D. Finn, J. Clements, S.R. Eddy Nucleic Acids Research (2011).

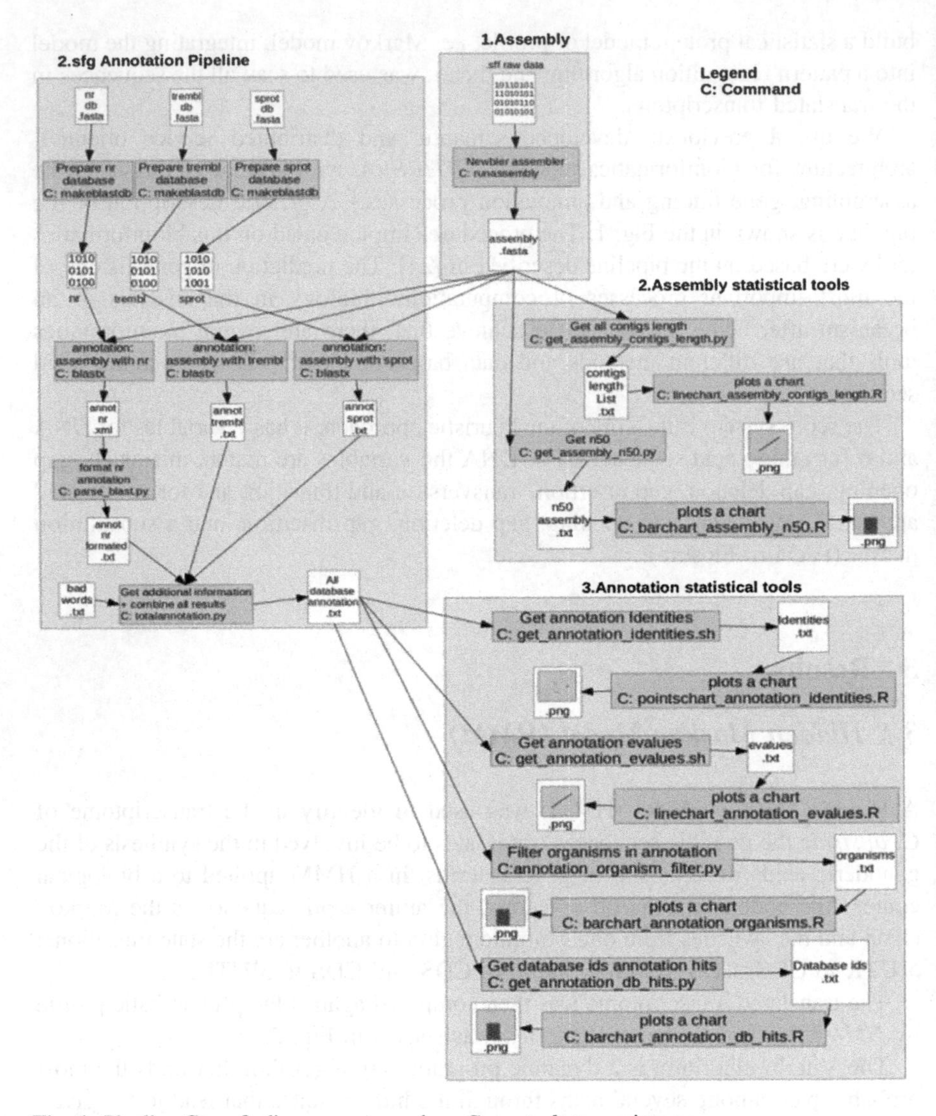


**Fig. 1** Pipeline Gene finding process used on *G. australe* transcriptome

purpose the bioinformatics tool *Transdecoder*[3] was used. Then, a probabilistic profile was built for the protein sequence of lanosterol synthase (EC 5.4.99.7), using conserved protein domains taken from the databases *pfamA*[4] and *tigrfam.*[5] Ten accessions under the Basidiomycota phyla (mushrooms, rusts, smuts, etc.) were used

[3]http://transdecoder.sourceforge.net/.

[4]ftp://ftp.sanger.ac.uk/pub/databases/Pfam/current_release.

[5]ftp://ftp.jcvi.org/pub/data/TIGRFAMs/.

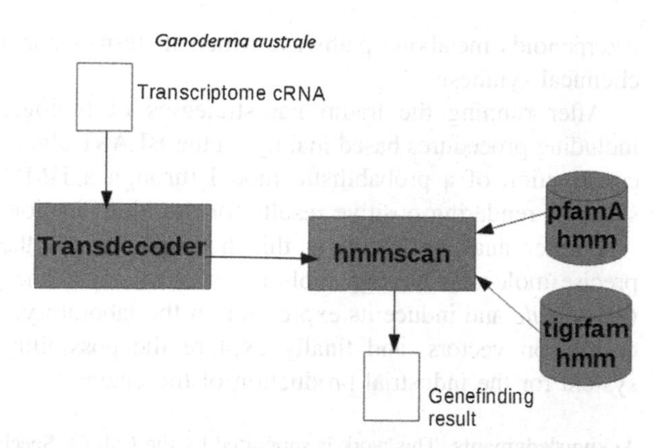


**Fig. 2** HMM Gene Finding pipeline

in order to maintain phylogenetic relationships as close as possible, A set of other 10 species under the Ascomycota phyla were used to complement de profile chosen as model organisms. In total, there were included 20 homologous sequences in the probabilistic model: Phyla (Division), Basidiomycota (10 species): Class Agaricomycetes, Order Polyporales, Family Ganodermataceae, *Ganoderma lucidum*; Class Agaricomycetes, Order Polyporales, Family, Polyporaceae, *Trametes versicolor* and *Dichomitus squalens*; Class Agaricomycetes, Order Agaricales, Family Psathyrellaceae, *Coprinopsis cinerea*; Class Agaricomycetes, Order Agaricales, Family Marasmiaceae, *Moniliophthora roreri*; Class Agaricomycetes, Order Auriculariales, Family Auriculariaceae, *Auricularia delicata*; Class Agaricomycetes, Order Cantharellales, Family Ceratobasidiaceae, *Rhizoctonia solani*; Subphylum Pucciniomycotina, Class Microbotryomycetes, Order Microbotryales, Family Microbotryaceae, *Microbotryum violaceum*; Subphylum Ustilaginomycotina, Class Ustilaginomycetes, Order Ustilaginales, Family Ustilaginaceae, *Pseudozyma brasiliensis*; Subdivision Agaricomycotina, Class Tremellomycetes, Order Tremellales, Family Tremellaceae, *Cryptococcus neoformans*. Phyla (Division), Ascomycota (10 species): *Neurospora crassa*, *Marssonina brunnea*, *Aspergillus niger*, *Aspergillus oryzae*, *Candida albicans*, *Candida maltosa*, *Fusarium oxysporum*, *Penicillium roqueforti*, *Saccharomyces cerevisiae* and *Trichoderma reesei*.

## 4 Conclusions

In this research we use a bioinformatics strategy for gene recognition based mainly on probabilistic methods through the hidden Markov models (HMM) theory, apply to the genome of a biotechnologically promising macromycete fungi, *Ganoderma australe*.

By this approach we were able to find significant evidence for the identification of the putative protein lanosterol synthase, involved in a key step in the triterpenoids metabolic pathway, which in turn is part of the ganoderic acid biochemical synthesis.

After running the traditional strategies of biological function identification, including procedures based mainly on the BLAST algorithm, it was evident that the construction of a probabilistic model through a HMM profile becomes a better strategy, rendering positive results for the identification of the putative sequence.

Further analysis based on this findings include the implementation of more precise molecular biology tools in order to isolate the gene from cell cultures of *G. australe* and induce its expression in the laboratory, with the use of appropriate expression vectors, and finally explore the possibility of a commercial scaling system for the industrial production of the enzyme.

**Acknowledgments** This work is supported by the Call for Special Project funding for Research and Innovation at the University of Caldas 2013.

## References

1. Punta, M., Ofran, Y.: The rough guide to in silico function prediction, or how to use sequence and structure information to predict protein function. PLoS Comput. Biol. **4**(10), e1000160 (2008). doi:10.1371/journal.pcbi.1000160
2. Watson, J.D., Laskowski, R.A., Thornton, J.M.: Predicting protein function from sequence and structural data. Curr. Opin. Struct. Biol. **15**, 275–284 (2005)
3. Lee, D., Redfern, O., Orengo, C.: Predicting protein function from sequence and structure. Nat. Rev. Mol. Cell Biol. **8**, 995–1005 (2007). doi:10.1038/nrm2281
4. Azad, R.K., Borodovsky, M.: Probabilistic methods of identifying genes in prokaryotic genomes: connections to the HMM theory. Briefings Bioinform. **5**(2), 118–130 (2004)
5. Erjavec, J., Kos, J., Ravnikar, M., Dreo, T., Sabotic, J.: Proteins of higher fungi – from forest to application. Trends Biotechnol. **30**(5), 259–273 (2012)
6. Rai, R.D., Singh, S.K., Yadav, M.C., Tewari, R.P.: Mushroom Biology and Biotechnology, Mushroom Society of India, Solan (H.P.) (2007)
7. Russell, R., Paterson, M.: Ganoderma – a therapeutic fungal biofactory. Phytochemistry **67**, 1985–2001 (2006)
8. Keller, N.P., Turner, G., Bennett, J.W.: Fungal secondary metabolism - from biochemistry to genomics. Nature Rev. Microbiol. **3**, 937–947 (2005)
9. Dinesh, B.P., Subhasree, R.S.: The sacred mushroom "Reishi"-a review. American-Eurasian J. Bot. **1**(3), 107–110 (2008)
10. Ozel, Y.K., Gedikli, S., Aytar, P., Unal, A., Yamac, M., Cabuk, A., Kolankaya, N.: New fungal biomasses for cyanide biodegradation. J. Biosci. Bioeng. **110**(4), 431–435 (2010)
11. Shimizu, E., Velez, J., Rueda, P., Zapata, L., Villalba, L.: Relación entre degradación de colorantes y oxidación de lignina residual causados por Ganoderma applanatum y Pycnoporus sanguineus en el licor negro kraft. Revista de Ciencia y Tecnología. **11**(12), 46–51 (2009)
12. Rigas, F., Papadopoulou, K., Dritsa, V., Doulia, D.: Bioremediation of a soil contaminated by lindane utilizing the fungus Ganoderma australe via response surface methodology. J. Hazard. Mater. **140**, 325–332 (2007)
13. Chuang, H.-W., Wang, I.-W., Lin, S.-Y., Chang, Y.-L.: Transcriptome analysis of cadmium response in Ganoderma lucidum. FEMS Microbiol. Lett. **293**, 205–213 (2009)

14. Chang, Y., Yang, J.S., Yang, J.L., Wu, C., Chang, S., Lu, K., Lin, J., Hsia, T., Lin, Y., Ho, C., Wood, W.G., Chung, J.: Ganoderma lucidum extracts inhibited leukemia WEHI-3 cells in BALB-c mice and promoted an immune response in vivo. Biosci. Biotechnol. Biochem. **73** (12), 2589–2594 (2009)
15. Karwa, A., Gaikwad, S., Rai, M.K.: Mycosynthesis of silver nanoparticles using Lingzhi or Reishi medicinal mushroom, Ganoderma lucidum (W. Curt.:Fr.) P. Karst and their role as antimicrobials and antibiotic activity enhancers. Int. J. Med. Mushrooms **13**(5), 483–491 (2011)
16. Li, N., Hu, Y.L., He, C.X., Hu, C.J., Zhou, J., Tang, G.P., Gao, J.Q.: Preparation, characterization, and anti-tumour activity of Ganoderma lucidum polysaccharide nanoparticles. J. Pharm. Pharmacol. **62**(1), 139–144 (2010)
17. González Muñoz, A., Botero Orozco, K.J., López Gartner, G.A.: Finding of a gene sequence related to the expression of a fungal immunomodulatory protein in Ganoderma australe. Revista Colombiana de Biotecnología, Universidad Nacional de Colombia **16**(2), 90–95 (2014). ISSN:0123-3475
18. Sliva, D., Labarrere, C., Slivova, V., Sedlak, M., Lloyd, F.P. Jr., Ho, N.W.Y.: Ganoderma lucidum suppresses motility of highly invasive breast and prostate cancer cells. Biochem. Biophys. Res. Commun. **298**, 603–612 (2002)
19. Yeh, C.H., et al.: Polysaccharides PS-G and protein LZ-8 from Reishi (Ganoderma lucidum) exhibit diverse functions in regulating murine macrophages and T lymphocytes. J. Agric. Food Chem. **58**, 8535–8544 (2010)
20. Wang, G., Zhao, J., Liu, J., Huang, Y., Zhong, J.-J., Tang, W.: Enhancement of IL-2 and IFN-γ expression and NK cells activity involved in the anti-tumor effect of ganoderic acid Me in vivo. Int. Immunopharmacol. **7**, 864–870 (2007)
21. Johnson, B.M., Doonan, B.P., Radwan, F.F., Haque, A.: Ganoderic acid dm: an alternative agent for the treatment of advanced prostate cancer. Open Prostate Cancer J. **3**, 78–85 (2010)
22. Ameri, A.: Ganoderic acid in the treatment of prostate cancer. Jundishapur J. Nat. Pharm. Prod. **7**(3), 85–86 (2012)
23. Shang, C.H., Shi, L., Ren, A., Qin, L., Zhao, M.W.: Molecular cloning, characterization, and differential expression of a lanosterol synthase gene from Ganoderma lucidum. Biosci. Biotechnol. Biochem. **74**(5), 974–978 (2010)
24. De Wit, P., Pespeni, M.H., Ladner, J.T., Barshis, D.J., Seneca, F., Jaris, H., Overgaard Therkildsen, N., Morikawa, M. Palumbi, S.R.: The simple fool's guide to population genomics via RNA-Seq: an introduction to high-throughput sequencing data analysis. Mol. Ecol. Resour. **12**, 1058–1067 (2012)

# A New Bioinformatic Pipeline to Address the Most Common Requirements in RNA-seq Data Analysis

**Osvaldo Graña, Miriam Rubio-Camarillo, Florentino Fdez-Riverola, David G. Pisano and Daniel Glez-Peña**

**Abstract** Many bioinformatic programs have been developed to analyze data from RNA-seq experiments. These programs are widely used and often included in computational pipelines. Nevertheless, there does not seem to be a precise definition of what constitutes a proper workflow for this kind of data. We present here a new workflow that takes into account the most common requirements for RNA-seq analysis, and that is implemented as an automatic pipeline to perform an efficient and complete evaluation.

**Keywords** RNA-seq · NGS · Pipeline · Transcriptomics

O. Graña · D.G. Pisano
Bioinformatics Unit, Structural Biology and BioComputing Programme,
Spanish National Cancer Research Centre (CNIO), 3rd Melchor Fernández Almagro St, 28029 Madrid, Spain
e-mail: ograna@cnio.es

D.G. Pisano
e-mail: dgpisano@cnio.es

M. Rubio-Camarillo
Structural Computational Biology Group, Structural Biology and BioComputing Programme,
Spanish National Cancer Research Centre (CNIO), 3rd Melchor Fernández Almagro St, 28029 Madrid, Spain
e-mail: mrubioc@cnio.es

F. Fdez-Riverola · D. Glez-Peña (✉)
ESEI - Escuela Superior de Ingeniería Informática Edificio Politécnico,
Campus Universitario as Lagoas S/N Universidad de Vigo, 32004 Ourense, Spain
e-mail: dgpena@uvigo.es

F. Fdez-Riverola
e-mail: riverola@uvigo.es



R. Overbeek et al. (eds.), *9th International Conference on Practical Applications of Computational Biology and Bioinformatics*, Advances in Intelligent Systems and Computing 375, DOI 10.1007/978-3-319-19776-0_13

## 1 Introduction

Sequencing of RNA molecules using next generation sequencing technologies (RNA-seq) has gained a lot of interest in the last years for transcriptomics [1].

RNA-seq is currently being applied in at least two different types of studies: in big projects with considerable funding and a large number of patients or samples [2–5], where RNA-seq is combined with other high-throughput techniques (ChIP-seq, BS-seq, variant calling, etc.), and, in individual experimental laboratories where a more limited number of samples or conditions are analyzed in the context of specific research.

For an experimental laboratory that performs RNA-seq experiments, and, more specifically, for the bioinformaticians that analyze the data, it is mandatory to perform an integrative analysis to answer common specific questions.

RNA-seq data analysis pipelines [6–12] usually make use of well known programs for the different parts of the analysis (aligners, gene expression quantifiers and differential expression programs). The performance of these programs has been evaluated [13–17], but the complete protocols followed by the pipelines still lack a proper definition to perform a more integrative analysis, to satisfy biologists' demands.

Based on the experience of the CNIO Bioinformatics Unit, due to the number of RNA-seq experiments analyzed, we have identified four different levels in the processing of these data: an initial level that checks for the quality of the sequenced reads and for possible sources of contamination. A second level of preprocessing that performs, if necessary, a trimming of low-quality nucleotides in the reads and/or a downsampling of reads to balance the different libraries. A third level to align the reads to the genome or transcriptome, and a fourth level where different analysis are done to obtain the results: assembly and quantification of transcripts, calibration of transcripts expression, differential expression between conditions, functional enrichment analysis, prediction of gene fusions, and finally, creation of bedgraph and bigwig files to upload to genome browsers. All together they facilitate a proper interpretation of the results.

After evaluating all the available pipelines for RNA-seq, we were not able to find even one that covered all the aspects described before. On top of that, it is necessary to provide human-readable output files to biologists to help in the interpretation of the results, because the direct files produced by these programs are usually raw text files that require a post-processing step, increasing the time needed for the analysis of each experiment, and leading to a lower efficiency and a decrease in productivity.

Due to these facts, we designed a new workflow that comprises the four levels described before (Fig. 1) and we implemented it as a new pipeline of quick and easy configuration and execution, in which all the steps are performed in an efficient manner, and where the manual processing is only reduced to the configuration side. The output files are properly decorated to allow for a better interpretation of the results.

To ensure a quick execution of the different levels in the workflow, the pipeline makes use of RUbioSeq libraries [18], which allow for a controlled execution in a single workstation, or, a parallelized execution in a computer cluster, adding scalability when the number of samples to process is higher.

## 2 Pipeline Overview

The pipeline is divided in four main levels: (1) Read quality and contamination checks, (2) read preprocessing through read trimming and/or down-sampling, (3) aligning of reads to the genomic or transcriptomic references and (4) processing of the derived alignment files to perform the different analysis (Fig. 1).

The pipeline execution can be a full execution of all the steps, or, only those that a user might want to execute or repeat. Input files are raw read files from the different samples, in FASTQ file format [19] or in unaligned binary SAM (BAM) format.

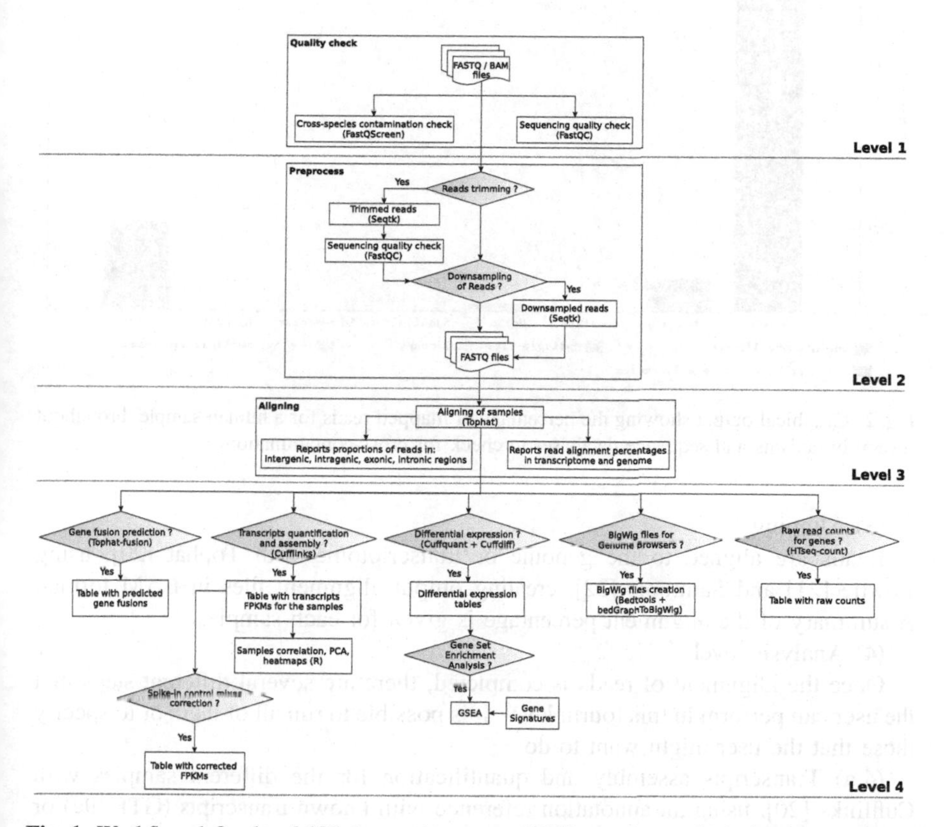


**Fig. 1** Workflow defined to fulfill the requirements of RNA-seq data analysis, implemented as a pipeline

Description of the different levels in the pipeline:

(1) Quality check: The quality of the sequenced reads is checked with FastQC (http://www.bioinformatics.babraham.ac.uk/projects/fastqc/). An output HTML file for each library is produced with the quality measures. At the same time, potential sources of contamination are checked with FastQScreen (http://www.bioinformatics.babraham.ac.uk/projects/fastq_screen/). A table and a graphical output are generated for each library (Fig. 2).

(2) Preprocessing: In case of detecting problems with reads quality in step 1, a user might choose to trim the reads from the start or the end, a task performed with Seqtk (https://github.com/lh3/seqtk). In this case, the read quality check from level 1 is repeated to ensure that the trimmed reads do not contain low-quality nucleotides. Also, libraries can be balanced regarding read number.

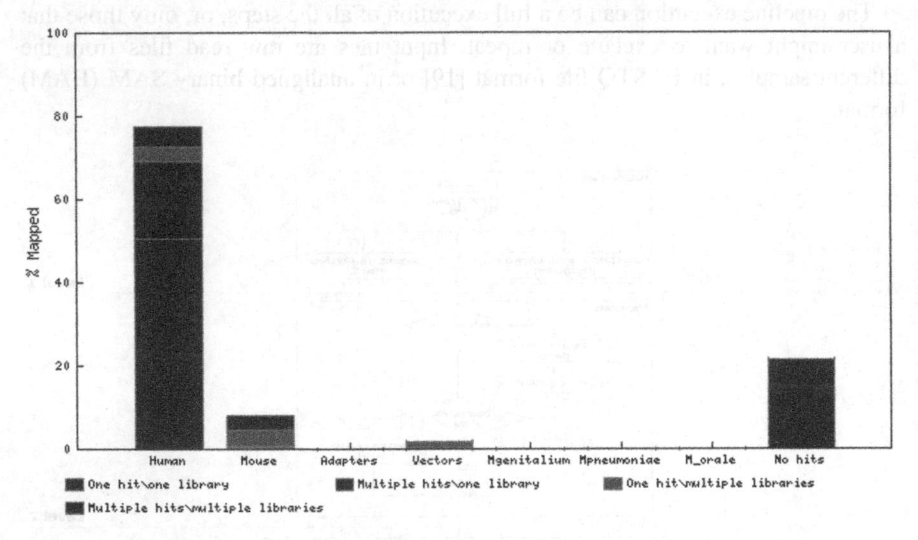


**Fig. 2** Graphical output showing the percentage of mapped reads for a human sample throughout several organisms and sequence databases to check for cross-contamination

(3) Aligning

Reads are aligned to the genome or transcriptome with Tophat [20], using Bowtie [21] and Samtools [22], creating output alignment files in BAM format. A summary of the alignment percentage is given for each sample.

(4) Analysis level

Once the alignment of reads is completed, there are several different steps that the user can perform in this fourth level. It is possible to run all of them or to specify those that the user might want to do:

(4.a) Transcripts assembly and quantification for the different samples with Cufflinks [20], using an annotation reference with known transcripts (GTF file) or without annotation reference. Once that the quantification of transcripts is done and to analyze the similarity among the different samples, the pipeline performs, by means of R functions, a Pearson correlation test and a principal component analysis (in 2D and 3D). The Pearson correlation test produces an output table with correlations among the samples (xls file) and the PCA results are shown with different pdf files (one 2D image and several 3D images from different angles to facilitate the view, Fig. 3). A table with the expression level of each gene across the samples is created in two different files, in xls and gct file formats (the latter allows the creation of heatmaps with GenePattern).

In case that the experiment is performed with spike-in controls, the pipeline can calibrate the expression level of the different transcripts using an R function [23], and producing a new xls table with the corresponding correction for the expression values.

Cuffmerge is also executed [20] to generate a new combined annotation among the initial reference annotation and the transcripts assembled from the samples in the experiment. This creates a new GTF file with an enriched annotation that could be used in the differential expression test.

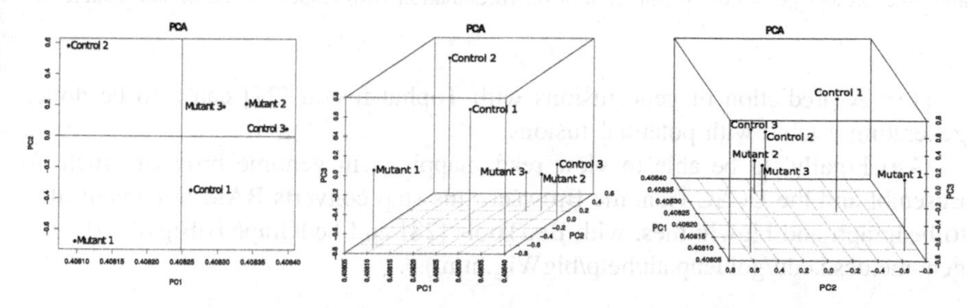


**Fig. 3** 2D and 3D PCAs produced by the pipeline, with different orientations, to facilitate the view

(4.b) Differential expression test with Cuffdiff [20] for the defined comparisons. The output files produced by Cuffdiff for differential expression for genes and gene isoforms are used to create two decorated and more human-readable xlsx files, with colors added to highlight the genes with the main differences regarding statistical significance: a first one with the full initial output, and a second one with only a subset containing the more reliable differences (Cuffdiff status OK and a expression value in FPKM > 0.05 in at least one of the conditions, Fig. 4).

Furthermore, the pipeline generates also a file in rnk format with a ranked list of genes. This list permits to perform a gene set enrichment analysis with GSEA [24] if one or more gene signatures are provided.

(4.c) It is possible to run HTSeq-count [25] within this level. A table with read counts for the genes across the different samples is generated. It can be used as a direct input to perform differential expression tests with DESeq and EdgeR [26].

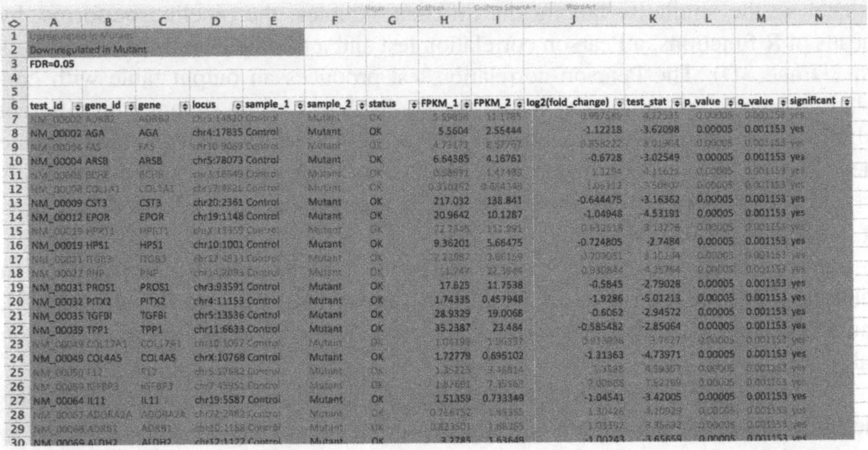

**Fig. 4** Screenshot showing a fragment of an output file created by the pipeline with differential expression values for gene isoforms. The yellow background highlights significant differences, red and green colors point out upregulation or downregulation with respect to the mutant condition

(4.d) A prediction of gene fusions with Tophat-fusion [27] can also be done, generating a table with potential fusions.

(4.e) Finally, to be able to view read mappings in genome browsers such as Ensembl and the UCSC Genome Browser, this step converts BAM alignment files to bedgraph and bigwig files, with BedTools [28] and bedGraphToBigWig (http://genome.ucsc.edu/goldenpath/help/bigWig.html).

## 2.1 Pipeline Execution

Only two XML files need to be configured to run the pipeline, once that all the programs that are required for each step are installed: (I) a configuration XML file, that defines the paths to the different programs and (II) an experiment XML file, that defines the samples and comparisons for each experiment (Fig. 5). Specific parameters for the different programs used in the analysis are specified here as well.

To speed up the execution of the different workflow levels, the pipeline makes use of the RUbioSeq parallelization libraries and allows two types of parallelization when run in a computer cluster: samples parallelization, allowing the processing of different samples at the same time, and analysis parallelization, permitting to launch different tasks of the same workflow level together. If the pipeline is executed in a single workstation, parallelization is allowed only for samples.

```
▾<experiment name="MDM2mutants" workspace="RNAseq/Analysis"
 referenceSequence="/Homo_sapiens/hg19/BowtieIndex/genome.fa"
 GTF="/Homo_sapiens/hg19/Annotation/genes.gtf"
 pairedEnd="false">
 ▾<library name="Control_1" leftFile="20140612.fastq">
    <rightFile/>
    <type>fastq</type>
    <solexaQualityEncoding/>
    <libraryType>firststrand</libraryType>
   ▾<trimming do="false">
      <nNucleotidesLeftEnd>3</nNucleotidesLeftEnd>
      <nNucleotidesRightEnd>5</nNucleotidesRightEnd>
    </trimming>
   ▾<downsampling do="false">
      <seed>3</seed>
      <nReads>100000</nReads>
    </downsampling>
    <mateInnerDist>150</mateInnerDist>
    <mateStdDev>50</mateStdDev>
  </library>
 ▾<library name="Mutant_1" leftFile="20140612.fastq">
    <rightFile/>
    <type>fastq</type>
    <solexaQualityEncoding/>
    <libraryType>firststrand</libraryType>
   ▾<trimming do="false">
      <nNucleotidesLeftEnd>3</nNucleotidesLeftEnd>
      <nNucleotidesRightEnd>5</nNucleotidesRightEnd>
    </trimming>
   ▾<downsampling do="false">
      <seed>3</seed>
      <nReads>100000</nReads>
    </downsampling>
    <mateInnerDist>150</mateInnerDist>
    <mateStdDev>50</mateStdDev>
  </library>
▾<comparison name="MutantVSControl">
 ▾<condition name="Control" cuffdiffPosition="1">
    <libraryName>Control_1</libraryName>
    <libraryName>Control_2</libraryName>
  </condition>
 ▾<condition name="Mutant" cuffdiffPosition="2">
    <libraryName>Mutant_1</libraryName>
    <libraryName>Mutant_2</libraryName>
  </condition>
 </comparison>
▾<tophat useGTF="true" nTophatThreads="4" maxMultihits="5"
 readMismatches="2" segmentLength="20" segmentMismatches="1"
 spliceMismatches="0" reportSecondaryAlignments="false"
 bowtie="1" readEditDist="2" readGapLength="2"
 referenceIndexing="false">
   <coverageSearch>--no-coverage-search</coverageSearch>
   <fusionSearchExperiment performFusionSearch="false">
   </fusionSearchExperiment>
 </tophat>
 <cufflinks nThreads="4" fragBiasCorrect="true"
 multiReadCorrect="true" libraryNormalizationMethod="classic-
 fpkm"></cufflinks>
 <cuffmerge nThreads="4" useCuffmergeAssembly="false">
 </cuffmerge>
 <cuffquant nThreads="4" fragBiasCorrect="true"
 multiReadCorrect="true" seed="123L"></cuffquant>
 <cuffnorm nThreads="4" outputFormat="simple-table"
 libraryNormalizationMethod="geometric" seed="123L"
 normalization="compatibleHits"></cuffnorm>
 <cuffdiff nThreads="4" fragBiasCorrect="true"
 multiReadCorrect="true" libraryNormalizationMethod="geometric"
 FDR="0.05" minAlignmentCount="10" seed="123L"
 FPKMthreshold="0.05"></cuffdiff>
▾<htseqcount minaqual="0" featuretype="exon" idattr="gene_id">
   <mode>intersection-nonempty</mode>
 </htseqcount>
```


**Fig. 5** Fragment of a XML file with the description of the samples, comparisons and specific parameters used by the programs executed by the pipeline, for a particular experiment

## 3 Conclusions

We define here a comprehensive workflow for the automatic analysis of RNA-seq data, implemented as a pipeline that is able to perform a number of tasks that together provide answers to the questions arose by biologists. The pipeline is implemented mainly in Perl, to facilitate the integration with RUbioSeq libraries. It also executes R functions to perform some small parts of the analysis. For each task in the different levels of the workflow, the pipeline makes use of state-of-the-art programs to produce consistent results. The CNIO Bioinformatics Unit and other CNIO laboratories have adopted this workflow as the standard protocol for RNA-seq data analysis. The software package is distributed under Creative Commons License. It can be freely downloaded (http://bioinfo.cnio.es/people/ograna/public_html/Software), installed and executed in Unix-based machines, like workstations or computer clusters with Linux or MAC OSX operative systems.

**Acknowledgements** This work was partially funded by the [14VI05] Contract-Programme from the University of Vigo. Also, it was supported by the European Union's Seventh Framework Programme FP7/REGPOT-2012-2013.1 under grant agreement n° 316265 (BIOCAPS), the Agrupamento INBIOMED from DXPCTSUG-FEDER "unha maneira de facer Europa" (2012/273) and the "Platform of integration of intelligent techniques for analysis of biomedical information" project (TIN2013 47153-C3-3-R) from the Spanish Ministry of Economy and Competitiveness.

## References

1. Wang, Z., Gerstein, M., Snyder, M.: RNA-Seq: a revolutionary tool for transcriptomics. Nat. Rev. Genet. **10**(1), 57–63 (2009)
2. International Cancer Genome Consortium, et al: International network of cancer genome projects. Nature **464**(7291), 993–998 (2010)
3. Abbott, A.: Europe to map the human epigenome. Nature **477**(7366), 518 (2011)
4. ENCODE Project Consortium: An integrated encyclopedia of DNA elements in the human genome. Nature **489**(7414), 57–74 (2012)
5. Cancer Genome Atlas Research Network et al.: The cancer genome atlas pan-cancer analysis project. Nat. Genet. **45**(10), 1113–1120 (2013)
6. Goncalves, A., Tikhonov, A., Brazma, A., Kapushesky, M.: A pipeline for RNA-seq data processing and quality assessment. Bioinformatics **27**(6), 867–869 (2011)
7. Goecks, J., Nekrutenko, A., Taylor, J.: Galaxy Team. Galaxy: a comprehensive approach for supporting accessible, reproducible, and transparent computational research in the life sciences. Genome Biol. **11**(8), R86 (2010)
8. Cumbie, J.S., Kimbrel, J.A., Di, Y., Schafer, D.W., Wilhelm, L.J., Fox, S.E., Sullivan, C.M., Curzon, A.D., Carrington, J.C., Mockler, T.C., Chang, J.H.: GENE-counter: a computational pipeline for the analysis of RNA-Seq data for gene expression differences. PLoS ONE **6**(10), e25279 (2011)
9. Reich, M., Liefeld, T., Gould, J., Lerner, J., Tamayo, P., Mesirov, J.P.: GenePattern 2.0. Nat. Genet. **38**(5), 500–501 (2006)
10. Knowles, D.G., Röder, M., Merkel, A., Guigó, R.: Grape RNA-Seq analysis pipeline environment. Bioinformatics **29**(5), 614–621 (2013)
11. Kalari, K.R., Nair, A.A., Bhavsar, J.D., O'Brien, D.R., Davila, J.I., Bockol, M.A., Nie, J., Tang, X., Baheti, S., Doughty, J.B., Middha, S., Sicotte, H., Thompson, A.E., Asmann, Y.W., Kocher, J.P.: MAP-RSeq: mayo analysis pipeline for RNA sequencing. BMC Bioinform. **15**, 224 (2014)
12. Torres-García, W., Zheng, S., Sivachenko, A., Vegesna, R., Wang, Q., Yao, R., Berger, M.F., Weinstein, J.N., Getz, G., Verhaak, R.G.: PRADA: pipeline for RNA sequencing data analysis. Bioinformatics **30**(15), 2224–2226 (2014)
13. Engström, P.G., Steijger, T., Sipos, B., Grant, G.R., Kahles, A., Rätsch, G., Goldman, N., Hubbard, T.J., Harrow, J., Guigó, R.: Bertone P; RGASP Consortium. Systematic evaluation of spliced alignment programs for RNA-seq data. Nat. Methods **10**(12), 1185–1191 (2013)
14. Soneson, C., Delorenzi, M.: A comparison of methods for differential expression analysis of RNA-seq data. BMC Bioinform. **14**, 91 (2013)
15. Rapaport, F., Khanin, R., Liang, Y., Pirun, M., Krek, A., Zumbo, P., Mason, C.E., Socci, N. D., Betel, D.: Comprehensive evaluation of differential gene expression analysis methods for RNA-seq data. Genome Biol. **14**(9), R95 (2013)
16. Steijger, T., Abril, J.F., Engström, P.G., Kokocinski, F., Hubbard, T.J., Guigó, R., Harrow, J., Bertone, P.: RGASP Consortium. Assessment of transcript reconstruction methods for RNA-seq. Nat. Methods **10**(12), 1177–1184 (2013)
17. Fonseca, N.A., Marioni, J., Brazma, A.: RNA-Seq gene profiling - A systematic empirical comparison. PLoS ONE **9**(9), e107026 (2014)
18. Rubio-Camarillo, M., Gómez-López, G., Fernández, J.M., Valencia, A., Pisano, D.G.: RUbioSeq: a suite of parallelized pipelines to automate exome variation and bisulfite-seq analyses. Bioinformatics **29**(13), 1687–1689 (2013)
19. Cock, P.J., Fields, C.J., Goto, N., Heuer, M.L., Rice, P.M.: The Sanger FASTQ file format for sequences with quality scores, and the Solexa/Illumina FASTQ variants. Nucl. Acids Res. **38**(6), 1767–1771 (2010)
20. Trapnell, C., et al.: Differential gene and transcript expression analysis of RNA-seq experiments with TopHat and Cufflinks. Nat. Protoc. **7**, 562–578 (2012)

21. Langmead, B., Trapnell, C., Pop, M., Salzberg, S.L.: Ultrafast and memory-efficient alignment of short DNA sequences to the human genome. Genome Biol. **10**(3), R25 (2009)
22. Li, H., et al.: The Sequence Alignment/Map format and SAMtools. Bioinformatics **25**(16), 2078–2079 (2009)
23. Lovén, J., Orlando, D.A., Sigova, A.A., Lin, C.Y., Rahl, P.B., Burge, C.B., Levens, D.L., Lee, T.I., Young, R.A.: Revisiting global gene expression analysis. Cell **151**(3), 476–482 (2012)
24. Subramanian, A., Tamayo, P., Mootha, V.K., Mukherjee, S., Ebert, B.L., Gillette, M.A., Paulovich, A., Pomeroy, S.L., Golub, T.R., Lander, E.S., Mesirov, J.P.: Gene set enrichment analysis: a knowledge based approach for interpreting genome-wide expression profiles. Proc. Natl. Acad. Sci. U S A **102**(43), 15545–15550 (2005)
25. Anders, S., Pyl, P.T., Huber, W.: HTSeq-a Python framework to work with high-throughput sequencing data. Bioinformatics **31**(2), 166–169 (2015)
26. Anders, S., McCarthy, D.J., Chen, Y., Okoniewski, M., Smyth, G.K., Huber, W., Robinson, M.D.: Count-based differential expression analysis of RNA sequencing data using R and Bioconductor. Nat. Protoc. **8**(9), 1765–1786 (2013)
27. Kim, D., Salzberg, S.L.: TopHat-Fusion: an algorithm for discovery of novel fusion transcripts. Genome Biol. **12**(8), R72 (2011)
28. Quinlan, A.R., Hall, I.M.: BEDTools: a flexible suite of utilities for comparing genomic features. Bioinformatics **26**(6), 841–842 (2010)

# Microarray Gene Expression Data Integration: An Application to Brain Tumor Grade Determination

**Eduardo Valente and Miguel Rocha**

**Abstract** World Health Organization ranks brain tumors in four stages, being the fourth grade the most aggressive. Glioblastoma, a fourth grade tumor, is one of the most severe human diseases that almost inevitability leads to death. Physicians address the classification in grades through direct inspection. Indeed, there is a need for good automatic predictors of tumor grade, which are not affected by human misclassification errors and that can be made with less invasive diagnostic tools. This work address the stages involved in the process of selecting a good tumor grade predictor, based on microarray gene expression data. In this work, the information integration from heterogeneous platforms is highlighted, evidencing the particularities of choosing approaches working at gene, transcript or probeset levels. Distinct machine learning algorithms and integration methods are tested, analyzing their ability to produce a good set of predictors for tumor grade.

**Keywords** Gene expression • Glioblastoma • Microarrays data integration • Filter methods • Wrapper methods

## 1 Introduction

The correct classification of the pathology of a patient, in particular in cancer, is essential to decide which drugs or therapies may be applied [1]. The diagnosis of cancer is usually done through microscopic and immunological tests over tissues biopsies, an extremely evasive process. Often, tumour samples with an atypical morphology complicate the analysis. In addition, certain types or subtypes of

E. Valente (✉)
Computer Science and Information Systems, IPCB, Castelo Branco, Portugal
e-mail: eduardo@ipcb.pt

M. Rocha
Centre of Biological Engineering, University of Minho, Braga, Portugal
e-mail: mrocha@di.uminho.pt



R. Overbeek et al. (eds.), *9th International Conference on Practical Applications of Computational Biology and Bioinformatics*, Advances in Intelligent Systems and Computing 375, DOI 10.1007/978-3-319-19776-0_14

tumours may have very little differentiation between them [2]. The analysis based on gene expression is extremely important, given the gaps that traditional diagnosis methods still present. These data are obtained by measuring the amounts of mRNA in a sample for the different genes in study. DNA microarrays can monitor simultaneously expression profiles from a large number of genes [3], as each microarray slide can carry a high amount of probes. Microarrays provide a comprehensive representation of gene expression levels for a given cell in just one experiment, for a given condition [4]. There are large amounts of gene expression data available on public databases, like GEO (Gene Expression Omnibus) and TCGA (The Cancer Genome Atlas), but each dataset comes from an independent experiment and has few available informative samples. The result of a statistical analysis on one of these datasets suffers from a lack of a good representative sample set. To improve the strength of these tests, it is necessary to integrate data from several sources, but this approach leads to complex challenges. The huge amounts of probeset readings lead to a requirement for feature filtering, until achieving a reduced set that could be used on traditional machine learning algorithms. Generally, the selection of the classification algorithm is determined by the researcher's familiarity with these algorithms, and not by the characteristics of the data or the algorithms themselves [2]. Any combination of feature set and classification algorithm should then be tested in new independent datasets to assess their strength and immunity to overfitting.
The different measurement processes also demands an integration method that overcomes that differences. In this work, data was collected from public databases at probeset level, from two microarrays vendors: Affymetrix® and Agilent®.

**Affymetrix Probeset Expression Measurement.** An Affymetrix probeset contains 11 or 16 25-mer probes, each probe measuring a specific zone of the gene. The small size of the sequence means that there is a great chance of cross-hybridization in these probesets. Each probe of Affymetrix is actually a pair of probes. One of these probes is the exact complementary sequence of the target to measure, called Perfect Match (PM), and the other has only a change of one base in the middle of the string, called Mismatch (MM) [5]. This change in the basis of MM probes is considered sufficient to not hybridize with the target, so it is considered that this probe measures background noise. The specific intensity for a probe is thus given by Eq. 1.

$$Probe\,Signal = PM - MM \tag{1}$$

To calculate the probeset expression, Affymetrix statistical methods are applied, which aim, among other purposes, to exclude from the calculation the pairs of probes that present levels outside the normal parameters [6]. The final value of a probeset measurement is often a log2 intensity (Eq. 2).

$$Probeset\,Signal = \log2(\text{probeset intensity}) \tag{2}$$

**Agilent Probeset Expression Measurement.** An Agilent probeset is composed of several identical probes for spot, with a target of 60-mers oligonucleotides. The measurement of the signal by spot is made using an image processing that measures the average intensity of pixels of this spot [7]. The datasets used on this work use a cell line, established through reference samples of various tissues, which serves to ameliorate the errors inherent in the platform, like for example the background error. The cell line is marked with green dye (Cy3) and the sample tissue is marked with red dye (Cy5). The signal that is returned is a logarithm of the ratio (Eq. 3).

$$Probeset\,Signal = log2(Cy5/Cy3) \tag{3}$$

Integrating multi-platform microarray data raises many difficulties, and therefore integration is often avoided due to the lack of good results. The authors who take this challenge focus on specific integration issues, and the methods used widely diverge. Kostadinova [8] examines how the combination of several related microarray datasets affects different areas of preprocessing and analysis of gene expression data, such as missing value imputation, gene clustering and biomarkers detection. Meng et al. [9] use a top-level approach, with multiple co-inertia analysis to relate two datasets, by maximizing the covariance between eigenvectors in paired dataset analysis. With this method, they can integrate and compare multi-omics data, independent of data annotation. The majority of the authors also focus on a single vendor or on datasets that are only one-channel or two-channel based. Papiez et al. [10] combined data sets from two types of microarrays: oligonucleotide and cDNA. They extract a set of genes common for both platforms and remove batch effects obtaining combined p-values from the two experiments. Even being different platforms, both datasets considered only intensity values. Although being difficult, data integration has been seen as an important achievement, even leading to proprietary solutions, like the ones described by Willis et al. [11]. In this work, a combination of Thomson Reuters solutions was used, including the Data Annotation and Processing tool, MetaCore Pathway Analysis and the Biomarkers Module of Thomson Reuters Integrity for a simple workflow to process omics data and identify biomarkers of target engagement through pathway analysis and data integration.

It is rare to find a work that integrates data from platforms with a different measurement basis, like the log intensities and log ratios that are approached in our work.

The aim of this work is to make a study of the features and mechanisms involved in the definition of a good prediction system through microarray data, for brain tumor grade classification. Based on some methods of data integration, feature selection and classification, it should be possible to develop analytical methods that are stable when faced with different sets of data and able to relate, in a robust way, statistical significance with biological relevance.

## 2 Methods

The problem to solve is a classification task, with 4 distinct classes, where grade 1 was excluded due to lack of representative samples on datasets and grade 0 refers to non-tumor. To exemplify, the distribution of the expression values across samples of dataset GDS1962 (Affymetrix) is shown in Fig. 1, specifically for the probeset 201292_at (gene TOP2A). This probeset was highlighted from a previous eBayes analysis, presenting a p-value of 8.02E-34 in this analysis, run to determine differential expression significance.
Four candidate algorithms were used, suitable for multi-level classification tasks: Ordinal Logistic Regression (OLR); Support Vector Machines (SVM); k-Nearest Neighbors (k-NN); and Linear Discriminant Analysis (LDA).

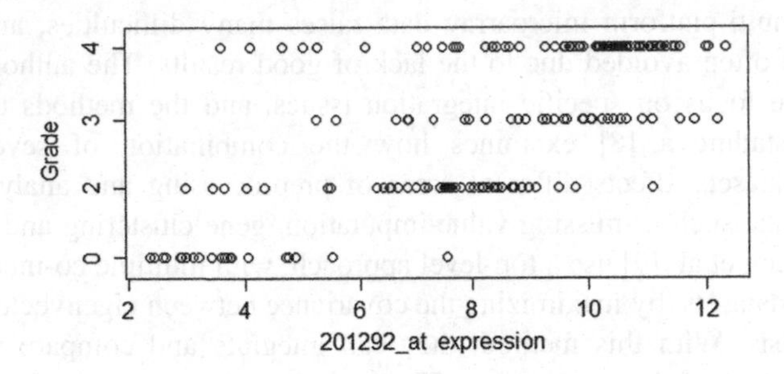


**Fig. 1** Expression/grade distribution for Affymetrix probeset 201292_at

The basic idea to choose the best pair of train algorithm - feature set is to choose the possible algorithm candidates and test each one against all possible combinations of feature sets. Obviously, this would result on a large processing burden. Then, the first step is to do a filtering stage, to reduce the initial set to a more reasonable subset. Two filtering methods will be used: Pearson's correlation of each feature with the tumor grade and empirical Bayes (eBayes) from R package *limma*, which will give a rank with p-values of the features that best differentiate the four tumor grades. Since some algorithms do not deal well with data collinearity, features that have a high Pearson's correlation among them will be eliminated. With the reduced feature set (rfSet), a greedy algorithm is deployed to choose, for each algorithm, the best feature subset:

```
bestFeatureCombinations ← NULL
setSize ← 1
while(setSize <= maxSetSize)
        newCombinations ← allCombinations(bestFeatureCombinations,
          rfSet)
        model ← trainAlgorithm(newCombinations)
        errorRate ← LOOCV(model)
        bestFeatureCombinations ← bestNewCombinations(errorRate)
        setSize ← setSize + 1
```

The first iteration is made with only sets with a single feature, testing each algorithm with each of the filtered features. The features which present the best results will be kept and then used as base for the next set of tests, where an additional feature, from the rest of filtered features, is included to organize sets of 2-features combinations. This method is repeated until reaching the *maxSetSize* threshold. The test criterion was the lower classification error rate (number of misclassifications over total number of samples) by Leave One Out Cross Validation (LOOCV). For OLR the Akaike's Information Criterion (AIC) was used, due to lower computation burden.

Firstly, the method will be applied separately to Affymetrix and Agilent datasets, to verify the strength of learning with relative homogeneous datasets. Next, the two types of platforms will be integrated and the same method will be applied to verify the differences. The integration will be made in six different ways:

- *Raw*, where the datasets are joined without any transformation;
- *Norm*, where the data is mean subtracted and divided by standard deviation;
- *Linear*. A set of reference features are selected, that present the lowest dataset and inter-grade expression variation in all datasets. When joining two datasets, one is used as base and a linear model is built that represents the linear variation between expression values of references from the base and the references of the new dataset. That model is then used to adjust the entire values of the new dataset. This is applied to linear data, without log2 transformation;
- *LinearLog2*. The same as the previous, but applied to log2 transformed data;
- *Grade*. Since the grade is an ordered value, dividing all data by the mean of values of common grade brings the heterogeneous datasets to levels that are comparable. This is applied to linear data, without log2 transformation;
- *GradeLog2*. The same as the previous but applied to log2 transformed data;

# 3 Results

For the tests, we selected datasets related with glioblastoma, where it was possible to distinguish tumor grade. Three datasets came from GEO, built with Affymetrix HG-U133Plus2.0, HG-U133A and HG-U133B platforms: GDS1962 (23 samples grade 0, 45 g2, 31 g3, 81 g4); GDS1815_6 (24 g3, 76 g4); and GDS1975_6 (26 g3, 59 g4). Two datasets were taken from TCGA, built with Agilent platforms: TCGA_AGIL2 (10 g0, 232 g4); and TCGA_AGIL4 (7 g2, 20 g3). As GDS1962 is the most complete and balanced dataset, it was used for preliminary Affymetrix probesets tests. For Agilent, TCGA_AGIL2 g4 subset was trimmed, leaving only 40 g4 samples. The two Agilent datasets were joined on TCGA_AGIL2_4 (10 g0, 7 g2, 20 g3, 40 g4). Higher-level datasets were built from probeset data. The gene level datasets were produced averaging all probesets associated with each gene. The transcript datasets were produced by weighted average of the probesets associated with each transcript. The two processes are described in detail in [12].

Pearson's correlation and eBayes filters were applied on these datasets and the probesets were sorted by rank. Over this, a cleaning process was performed to remove collinear probesets, using a 0.9 Pearson's correlation cutoff. Finally, only a set of the 200 best ranked probesets were kept for further tests.

The wrapper method was tuned to make combinations until a maximum of 20 features. Also, the quantity of the best combinations preserved between cycle iteration was limited to 20. For k-NN, k was always set to 3. The main results are provided in Tables 1, 2 and 3.

**Table 1** Results for classifiers (columns), filtering methods (rows) at the probeset level. #P - number of probesets, ER - Error Rate

| | | OLR | SVM | k-NN | LDA |
|---|---|---|---|---|---|
| Pearson correlation | Affy | #P: 16 ER: 0.25 | #P: 13 ER: 0.11 | #P: 18 ER: 0.14 | #P: 13 ER: 0.19 |
| | Agil | #P: 02 ER: 0.13 | #P: 6 ER: 0.04 | #P: 06 ER: 0.01 | #P: 05 ER: 0.04 |
| eBayes | Affy | #P: 16 ER: 0.28 | #P: 10 ER: 0.09 | #P: 20 ER: 0.14 | #P: 12 ER: 0.17 |
| | Agil | #P: 05 ER: 0.17 | #P: 09 ER: 0.03 | #P: 06 ER: 0 | #P: 04 ER: 0.03 |

**Table 2** Results for classifiers (*columns*), filtering methods (*rows*) at the gene level. #G - number of genes, ER - Error rate

| | | OLR | SVM | k-NN | LDA |
|---|---|---|---|---|---|
| Pearson Correlation | Affy | #G: 16 ER: 0.26 | #G: 19 ER: 0.08 | #G: 14 ER: 0.14 | #G: 19 ER: 0.17 |
| | Agil | #G: 03 ER: 0.12 | #G: 08 ER: 0.03 | #G: 05 ER: 0.01 | #G: 11 ER: 0 |
| eBayes | Affy | #G: 16 ER: 0.23 | #G: 17 ER: 0.09 | #G: 18 ER: 0.14 | #G: 19 ER: 0.17 |
| | Agil | #G: 05 ER: 0.17 | #G: 04 ER: 0.01 | #G: 06 ER: 0 | #G: 13 ER: 0 |

**Table 3** Results for classifiers (*columns*), filtering methods (*rows*) at the transcript level. #T - number of transcripts, ER - Error rate

| | | OLR | SVM | k-NN | LDA |
|---|---|---|---|---|---|
| Pearson Correlation | Affy | #T: 18 ER: 0.25 | #T: 14 ER: 0.1 | #T: 20 ER: 0.14 | #T: 17 ER: 0.17 |
| | Agil | #T: 03 ER: 0.09 | #T: 06 ER: 0.03 | #T: 07 ER: 0.01 | #T: 17 ER: 0.04 |
| eBayes | Affy | #T: 18 ER: 0.23 | #T: 18 ER: 0.07 | #T: 17 ER: 0.13 | #T: 13 ER: 0.16 |
| | Agil | not converge | #T: 06 ER: 0.02 | #T: 04 ER: 0.02 | #T: 06 ER: 0 |

The differences between results associated with the filtering method are not significant, so one will be used arbitrarily. It is also possible to see that the accuracy is always better for Agilent datasets, and SVMs have the best performance when considered both platforms. In the matters of probeset, gene or transcript level, the former presented slightly worse results. Between gene and transcript, it is not possible to distinguish a clear difference. Besides the identification of the best

classification algorithm, it is important to compare the prediction features obtained for Affymetrix and Agilent.
If a transcript/gene is related to the grade phenotype, then it should be relevant for both platforms. To check this premise, eBayes was applied to rank features separately, and then pairwise Kendall's coefficient of concordance (W) was calculated. For genes we obtained W = 0.675, and for transcripts W = 0.711. There is a considerable discordance about feature relevance between Affymetrix and Agilent. Deepening the analysis, the features selected as the best predictor with SVM for Affymetrix was used for Agilent tests and vice versa. With Affymetrix predictors, a LOOCV error of 0.10 for genes and 0.09 for transcripts, was obtained when these predictors were applied on Agilent. For Agilent predictors, an error of 0.29 for genes and 0.32 for transcripts was obtained, when applied on Affymetrix data. It is clear that predictors obtained from Agilent data tend to overfit more and loose generalization.
The next tests were made joining all datasets on a single one, using different methods for data integration (Table 4). SVMs are confirmed as the training algorithm that gives the best results and the division of the data by a reference set of values (grade) also improves accuracy. As the common grade between all datasets is g3, this was the reference grade used.
Joining heterogeneous datasets aims to produce a robust training set that allows to classify a new sample with a good rate of assertiveness. With these LOOCV tests it is possible to preliminary confirm that it is possible to do this with datasets from Agilent and Affymetrix.

**Table 4** LOOCV error rate for affymetrix and agilent data integration

| | | Raw | Norm | Linear | LinearLog2 | Grade | GradeLog2 |
|---|---|---|---|---|---|---|---|
| OLR | Genes | 0.24 | 0.28 | 0.27 | 0.31 | 0.2 | 0.2 |
| | Transcripts | 0.24 | 0.3 | 0.3 | 0.26 | 0.19 | 0.23 |
| SVM | Genes | 0.11 | 0.09 | 0.11 | 0.09 | 0.08 | 0.12 |
| | Transcripts | 0.09 | 0.1 | 0.12 | 0.15 | 0.08 | 0.11 |
| k-NN | Genes | 0.13 | 0.16 | 0.19 | 0.15 | 0.14 | 0.17 |
| | Transcripts | 0.11 | 0.18 | 0.19 | 0.2 | 0.14 | 0.17 |
| LDA | Genes | 0.16 | 0.2 | 0.19 | 0.16 | 0.14 | 0.18 |
| | Transcripts | 0.16 | 0.19 | 0.22 | 0.19 | 0.15 | 0.18 |

To take the tests a little further, the same joined dataset was used to build a model to predict a new Affymetrix dataset, collected from TCGA, not used before in any stage. This new dataset has 10 g0 and 287 g4 samples so, to apply the reference grade method, all data was divided by grade 0. The results again confirmed the good performance of the method (Table 5).

**Table 5** Confusion matrices for prediction of the new dataset

**Genes**

| | | Predicted | | | |
|---|---|---|---|---|---|
| | **grade** | 0 | 2 | 3 | 4 |
| Observed | 0 | 10 | 0 | 0 | 0 |
| | 4 | 0 | 6 | 5 | 266 |
| **error rate:** 0.038 | | | | | |

**Transcripts**

| | | Predicted | | | |
|---|---|---|---|---|---|
| | **grade** | 0 | 2 | 3 | 4 |
| Observed | 0 | 10 | 0 | 0 | 0 |
| | 4 | 0 | 4 | 8 | 265 |
| **error rate:** 0.042 | | | | | |

## 4 Conclusion

Glioblastoma is a disease that urgently needs new prognostic methods that provide an extra aid to the actual efforts. Microarrays have brought light to evidences connecting gene expression and phenotypes. However, existing microarray manufacturers do not adopt common standards, given as a result independent datasets only interpretable when used with extra information from vendors, like the probeset-gene mapping.

In this work, we step through the process of choosing the best set of brain tumor grade predictors and the best algorithms to train them. The filter methods reduce drastically the initial dimension of the datasets, and did not present great differences among them. SVMs stood out as the best classifiers for this case. The use of a reference grade to bring data to equivalent scales between datasets also shown to be a good method of integration. The level of data to use on the integration, gene or transcript, needs a more profound interpretation. The probesets used for expression measurement are exon centric and do not read all exons of a gene. Since the same exon could be part of several different transcripts, it is common to find many of them with the same expression value, which causes entropy to learning algorithms. Even filtering the 'clones', in many cases the expression value of a transcript is the same as the correspondent gene, when all the probesets measuring the gene also measure that transcript. It is than clear that it is necessary to upgrade the expression measurement method to be possible to use transcript levels effectively.

The final model has shown good results when applied to a new dataset.

**Acknowledgments** The work is partially funded by Project 23060, PEM - Technological Support Platform for Metabolic Engineering, co- funded by FEDER through Portuguese QREN under the scope of the Technological Research and Development Incentive system, North Operational.

## References

1. Cho, S., Ryu, J.: Classifying gene expression data of cancer using classifier ensemble with mutually exclusive features. Proc. IEEE **90**(11), 1744–1753 (2002)
2. Lorena, A., Costa, I., Souto, M.: On the complexity of gene expression classification data sets. In: 8th International Conference on Hybrid Intelligent Systems, pp. 825–830 (2008)
3. Huang, J., Fang, H., Fan, X.: Decision forest for classification of gene expression data. Comput. Biol. Med. **40**(8), 698–704 (2010)
4. Kuçukural, A., Yeniterzi, R., Yeniterzi, S., Sezerman, O.: Evolutionary selection of minimum number of features for classification of gene expression data using genetic algorithms. In: Proceedings of the 9th Annual confernce, GECCO 9, pp. 401–406 (2007)
5. Ballester, B., Johnson, N., Proctor, G., Flicek, P.: Consistent annotation of gene expression arrays. BMC Genom. **11**, 294–308 (2010)
6. Rouchka, E., Phatak, A., Singh, A.: Effect of single nucleotide polymorphisms on Affymetrix match-mismatch probe pairs. Bioinformation **2**(9), 405–411 (2008)
7. Zahurak, M., Parmigiani, G., Yu, W., Scharpf, R., Berman, D., Schaeffe, E., Shabbeer, S., Cope, L.: Pre-processing Agilent microarray data. BMC Bioinform. **8**(142) (2007)
8. Kostadinova, E.: Data Integration: an approach to improve the preprocessing and analysis of gene expression data. Union Sci Bulgaria **6**, 120–133 (2013)
9. Meng, C., Kuster, B., Culhane, A., Gholami, A.: A multivariate approach to the integration of multi-omics datasets. BMC Bioinformatics **15**, 162–175 (2014)
10. Papiez, A., Finnon, P., Badie, C., Bouffler, S., Polanska, J.: Integrating expression data from different microarray platforms in search of biomarkers of radiosensitivity. IWBBIO, Granada **1**, 484–493 (2014)
11. Willis, C., Whyte, K., Baker, M., Pestell, R.: Integrating transcriptomic data using metacore pathway analysis to identify novel biomarkers of bevacizumab target engagement. In: Reuters, T. (ed.) American Association for Cancer Research, San Diego, vol. 74 (2014)
12. Valente, E., Rocha, M.: (in press) Transcript-based reannotation for microarray probesets. In: ACM Symposium on Applied Computing, Salamanca (2015)

# Obtaining Relevant Genes by Analysis of Expression Arrays with a Multi-agent System

**Alfonso González, Juan Ramos, Juan F. De Paz and Juan M. Corchado**

**Abstract** Triple negative breast cancer (TNBC) is an aggressive form of breast cancer. Despite treatment with chemotherapy, relapses are frequent and response to these treatments is not the same in younger women as in older women. Therefore, the identification of genes that provoke this disease is required, as well as the identification of therapeutic targets. There are currently different hybridization techniques, such as expression arrays, which measure the signal expression of both the genomic and transcriptomic levels of thousands of genes of a given sample. Probesets of Gene 1.0 ST GeneChip arrays provide the ultimate genome transcript coverage, providing a measurement of the expression level of the sample. This paper proposes a multi-agent system to manage information of expression arrays, with the goal of providing an intuitive system that is also extensible to analyze and interpret the results. The roles of agent integrate different types of techniques, from statistical and data mining techniques that select a set of genes, to search techniques that find pathways in which such genes participate, and information extraction techniques that apply a CBR system to check if these genes are involved in the disease.

**Keywords** Expression arrays · Multi-agent system · CBR system · Pathway

A. González (✉) · J. Ramos · J.F. De Paz · J.M. Corchado
Biomedical Research Institute of Salamanca/BISITE Research Group,
University of Salamanca, Edificio I+D+I, 37008 Salamanca, Spain
e-mail: alfonsogb@usal.es

J. Ramos
e-mail: juanrg@usal.es

J.F. De Paz
e-mail: fcofds@usal.es

J.M. Corchado
e-mail: corchado@usal.es



R. Overbeek et al. (eds.), *9th International Conference on Practical Applications of Computational Biology and Bioinformatics*, Advances in Intelligent Systems and Computing 375, DOI 10.1007/978-3-319-19776-0_15

# 1 Introduction

There are several techniques that can be used to study genetic variation in patients, such as tissue microarrays, expression arrays (RNA) [4, 5], genomic arrays (DNA) and arrays of microRNAs (miRNAs). Arrays used in the study of expression profiling are cDNA arrays and oligonucleotide chips. Moreover, different types of genomic arrays (DNA) are used, including BAC aCGH, oligo CGH, SNP CGH and aCGH (Comparative Genomic Hybridization) [24]. CGH arrays (aCGH) can compare the DNA of a patient with control DNA and use this information to detect mutations [16, 20] based on the increase, loss or amplifications [23] in different regions of the chromosome. There are new exon arrays that provide accurate assessments of gene expression [12]. Information sources are varied but laboratory personnel usually follow fixed analysis processes that are distributed in sequences, which are in turn executed repeatedly in the search for genes that are considered to be relevant. Therefore, it is necessary to find a system that automates this process so that the work of the laboratory staff is simplified.

There have also been efforts to provide a solution to the main challenges associated with analyzing microarray data, which are: the high amount of data (coming from thousands of genes extracted from few samples); the high complexity of the data; the fact that gene datasets in microarrays are often correlated (either directly or indirectly); and the fact that most gene selection and prediction models emphasize the capacity for effective classification instead of the function of an effective selection. The assumption is that statistical significance is equivalent to biological importance.

There are other investigations which focus their efforts in predicting genes that cause diseases. Thanh-Phuong Nguyne and Tu-bao Ho have developed a semi-supervised framework in order to find genes and detect possible connections among those that can lead to those diseases [17]. They are based on feature extraction, preprocessing of data and integrate the following resources: Universal ProteinResource (UniProt) [21], Gene Ontology (GO) [10], Pfam [9], InterDom [18], Reactome [11] and to expression dataset [7]. Maglietta et al. [15] propose a method from a similar point of view. The target is the selection of genes relevant to a pathology by analyzing the tissue expression profiles for two different phenotypic conditions. Statistical techniques are used and the presence of genes in similar studies is verified. Other studies use multiagent systems in order to analyze array data, including a system proposed by Juan F. De Paz et al. [6], where a multiagent system analyses CGH arrays searching for gene gains or losses, which are then represented. This study is more oriented to obtain relevance areas and provide easy access to information but works only with CGH arrays. The present paper proposes a multi-agent system to analyze expression arrays. The main novelty is that the system can learn analysis flows (workflow) while the expression analysis is being performed, thus automating the analysis of expression. During the analysis, services are incorporated in order to carry out the analysis and extraction of information from database, through which most relevant genes are selected. Different data mining techniques and databases were used to analyze expression profiles and obtain relevant genes for two different phenotypic conditions. The system was applied to a real case study for the analysis of breast cancer with the aim of analyzing differences in this type of cancer with specific regard to the patient's age.

This article is organized as follows: Sect. 2 describes the state of the art of expression arrays, Sect. 3 describes the proposal, and Sect. 4 presents the results and conclusions.

## 2 Gene Expression Arrays

Microarrays constitute a widely used tool that measure gene expression [19]. Moreover, this technology has attracted a special interest in cancer research [13]. An expression arrays analysis makes it possible to study and compare transcriptomes of different samples. The value of gene expressions in these biochips is determined by the intensity of the hybridization of transcripts with a group of probes [12].

With these qualities, expression arrays become a very useful tool that makes it possible to determine which genes have an altered expression, to compare expressions based on certain parameters, and to diagnose and distinguish subtypes of cancers with similar clinical manifestations, among other things. Different kinds of cancer genes share groups and altered pathways. Array analysis can investigate typical genes, as well as those that are not common to the vast majority of proliferative syndromes [19], existing in more specific forms of the disease. This is one factor that makes arrays a useful diagnostic tool.

Beyond studying the expression of each gene and its degree of responsibility in an alteration, it is vital to understand the expression of these genes and the proteins they encode in the context of signaling pathways [25].

To be able to perform a complete analysis, one of the roles of a multi-agent system is to search in different databases for the pathways taken by the specific genes that are being studied. One mutation in a particular gene can give rise to various effects, even in the same type of tissue [22]. Because of this, the function of the platform is interesting, specifically because the information obtained from the study of a single gene is not representative if, after it has been studied, its relationship to other elements that also influence the signaling pathway is not verified [25].

The main function of this platform is to be able to select the relevant genes for the investigation. There comes a point during the screening process when there is no longer a sufficient number of elements to obtain pathways. It is precisely for this reason, from a research point of view, that it is important to compare the genes obtained from the analysis that best explain the gene alteration (depending on the studied parameter) according to the altered pathways.

Among the most influential resulting gene expression analyses of patient samples in our case study, those that are also therapeutic targets are of particular interest to medical research. One of the great difficulties of the analysis of arrays is to obtain biologically valid conclusions from vast amounts of data [14]. Consequently, one agent from the multi-agent system is responsible for conducting searches in databases known to contain therapeutic targets. This opens up the possibility of attacking these targets with drugs and conducting pharmacogenomic research after the analysis [19]. This makes it possible to check and directly study the influence of the pathway in the carcinogenic process, in addition to its clinical implications and the search for effective treatments.

## 3 Multi-agent System

In these studies, users have to work with a large volume of information, which involves the development of programs to improve data analysis systems and to automatically extract information through databases [3].

Our study uses expression arrays that determine the expression of genes to the probes used. This information is taken into account to observe differences that may occur in the same genes with regard to the age factor. Because large amounts of data are handled, it is necessary to develop a system aimed at simplifying the management and analysis of this information, and at automatically extracting information to determine the correlation of these genes in breast cancer.

Distributed analysis of expression data is performed by various personnel of the laboratory: from chip hybridization to the removal of variations and relevant information associated with the chips. This study shows a multiagent system specifically designed, with an abstract architecture for this virtual organizations [1], to analyze expression arrays. The functionality of the multi-agent system is divided into layers and roles to perform the analysis, which usually consists of several stages. The first stage is pre-processing, which performs the important task of the screening the data for the first time. The next stage performs an analysis of the expression probes, searching for differences with respect to the expression under normal conditions for that gene, or with respect to any specific factor. In the next stage data mining techniques are applied, allowing the data set studied to be further reduced. When looking for differences between groups of patients, it is important to confirm whether a cluster has been properly formed at the end of this stage, according to the case study. If a suitable result is not obtained, it will be necessary to review the previous step of extracting relevant genes through the data mining techniques. The final stage is initiated once the data set containing the different genes has been identified. This data set will be transferred to a database that checks the implication of these genes in the specific disease being studied to determine whether there is or is not a relationship.

JADE (Java Agent DEvelopment Framework) was adopted for the design and implementation of the proposed intelligent multi-agent and the architecture is composed of four layers: Analysis, Information Management, Visualization and Workflow. The architecture is shown in Fig. 1.

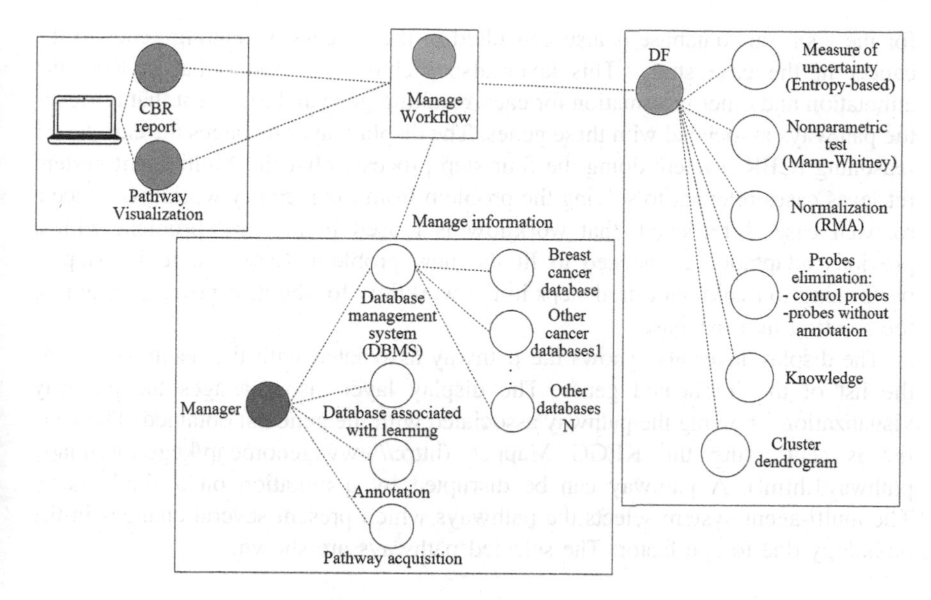


**Fig. 1** Multi-agent system architecture

The workflow layer includes an agent in charge of workflow in the other layers, and of establishing the correct order for the activity of each agent. Workflow analysis collects information about the settings and can repeat the sequences performed above for expression analysis. This aspect makes it possible to automate repetitive analysis tasks for laboratory personnel.

The analysis layer performs microarray analysis tasks as required by the process. This layer consists of several agents that are responsible for implementing the necessary processes and algorithms. An agent apply the Entropy-based filters, which is responsible for finding discrete attribute weights based on their correlation with continuous class attributes. Another agent applies the nonparametric Mann-Whitney statistical test to two independent samples. One more agent applies a normalization to adjust the signal, which may contain errors caused by technical and/or biological factors. The normalization technique applied is RMA (Robust multiarray Average), which adjusts information on each probeset to a comparable value. If any is over-expressed or under-expressed its genetic value is regulated with regard to the total. Another role performed by a system agent is the elimination of control probes and probes without notation. Finally, another agent produces a dendrogram that allows us to organize data into subcategories in which the case study is performed, this representation enables a clear representation of the relationship between data grouping.

The information management layer is responsible for confirming whether the genes obtained from the results of the analysis layer are related to the type of cancer that are the focus in the case study. This layer creates a database that collects the learned genes that are related to a cancer that is not contained in the databases used for the visit; this database is also consulted in the process of relating genes to the cancer in the case study. This layer also includes an Agent that collects the annotation and other information for each resulting gene and the agent that retrieves the pathways associated with these genes. The display layer manages the case based reasoning (CBR) system doing the four-step process. First the Multi-agent system retrieves cases relevant to solving the problem from the memory workflow. Once a relevant case is retrieved, that workflow is reused in the new problem with a previous adaptation as needed to fit the new problem. Later the revise step is performing to avoid undesired steps in the workflow for the next process: retaining the solution in a database.

The display layer also shows the pathway associated with the result as well as the list of the implicated genes. The display layer also manages the pathway visualization, showing the pathway associated with the gene list obtained. The gene list is sent using the KEGG Mapper (http://www.genome.jp/kegg/tool/map_pathway1.html). A pathway can be disrupted by a mutation on a single gene. The multi-agent system selects the pathways which present several changes in the pathology due to age factor. The selected pathways are shown.

## 4 Results

The case study was performed with 16 samples from patients with triple negative breast cancer provided by the Salamanca Cancer Institute. 8 samples corresponded to younger patients (less than 45 years) and the remaining 8 samples to elderly patients (over 68 years). Additional samples continue to be gathered in order to improve the accuracy of the results.

The technology used to analyze the microarray was Affymetrix, which is based on oligonucleotide chips. The specific chip used was the HuGene-1.0st-v1 chip, which contains 33,297 probes that identify about 23,000 sequenced genes.

The multi-agent system designed in this paper is applied to study gene expression arrays from the samples of these patients. The goal is to obtain the genes that show differences between samples from younger patients and older patients in order to discover why older women respond better to the treatment.

The first step in the analysis of oligonucleotide chips is the process of discarding any control probes and probes without notation. Once these probes are discarded a denormalization process [2] is performed. This process discards the values that deviate from the normal value, and applies the statistic test (Mann-Whitney) to each of the independent samples used in the case study (younger women and older women). In that step a Benjamini and Hochberg FDR multiple test correction is performed and apply on the p-value calculated based on the two-samples Mann-Whitney.

In our case of study, we look for variations that may occur in the expression levels of genes for samples associated with younger women compared with those of the older women. Once applied, we discard all values exceeding a p-value greater than 0.01, i.e., we keep the probes that have a greater interest because of their expression level compared with older youth at statistical level.

In this process of data analysis, once the preprocessing, normalization and application of statistical tests and techniques of data mining are completed, a clustering algorithm is applied. This algorithm provides us with a dendrogram which allows us to check the degree of clustering of the probes with respect to the two samples (probes for younger and older women). Process results are shown in Fig. 2.

The next stage is responsible for managing information received for the data obtained at the end of the analysis stage. These genes are contrasted with the corresponding type of cancer data base with which the data are associated (in this case study, the data base is for breast cancer). With this process we discard genes that are not implicated in a certain cancer, focusing on those that are.

The pertinent notation and associated pathways are obtained for this final gene set, allowing the access and use of this information.

The workflow layer agent learns the execution sequence of tasks, the order in which the agents interact with each other to execute the various processes and algorithms. With this initial learning, the following analysis of expression arrays are performed automatically, so the lab technicians do not have to perform the process manually, which avoids the risk of human error, loss of time, or a lower yield.

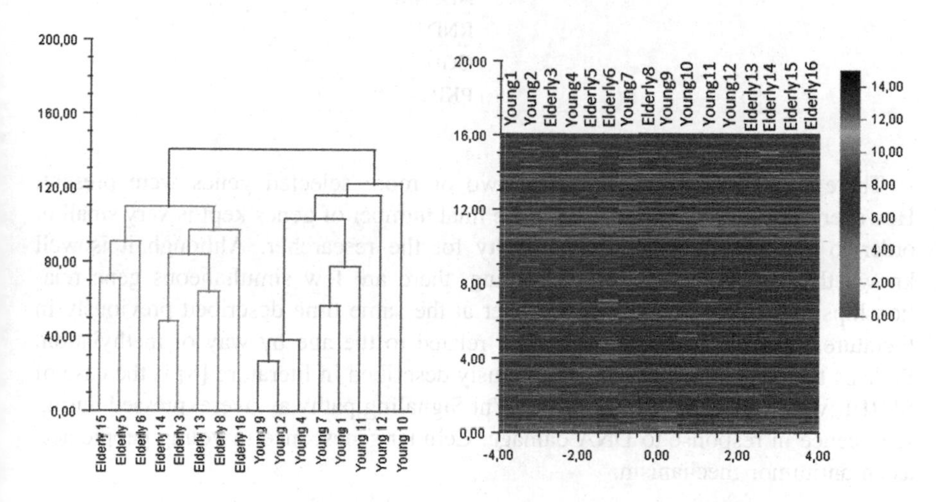


**Fig. 2** Dendogram and heatmap resulting

Once the genes and pathways are shown, an agent from the visualization layer performs a reasoning cycle (CBR). During the recovery step, information from the catalogued gene is obtained from the accurate database G2SBC (Genes-to-Systems Breast Cancer Database - http://www.itb.cnr.it/breastcancer/). While contrasting genes with the breast cancer database, these genes are evaluated according to the contrast hypothesis mentioned in part 3. In this way, if our system detects genes which are not in the databases and influence in the pathology according to the results, those genes are stored in an own database for future analysis.

Table 1 shows the most important genes obtained in the case study for triple negative breast cancer. These genes are considered by the system as the most important in the existence of differences in response to treatment of younger women versus olderer women.

**Table 1** Overexpressed genes resulting

| **Altered genes with close relationship to the age factor** | **Altered genes with no-apparent relation to the age factor** |
|---|---|
| GABRP | CAPN6 |
| SFRP1 | SLC6A14 |
| MID1 | SCRG1 |
| RARB | BCL11A |
| ACTG2 | PTGS2 |
| | BBOX1 |
| | S100A7 |
| | PRKAA2 |
| | ACPP |
| | ALCAM |
| | RND3 |
| | GGH |
| | PKP2 |

There was no pathway in which two or more selected genes were present. However, this is not surprising since the final number of genes kept is very small in order to provide a manageable quantity for the researcher. Although it is well known that expression varies with aging, there are few simultaneous gene relationships between age and breast cancer at the same time described previously in literature. Maybe cases like RARB are related to the age by way of methylation. Perhaps the clearest relationship previously described in literature [8] is the case of SFRP1, which is an antagonist of the Wnt Signaling pathway overexpressed during senescence in response to DNA damage. Cellular senescence in young people acts as an antitumor mechanism.

## 5 Conclusions

The developed system enables using patient samples to know if there are differences in the expression level for the proposed gene sets, allowing the system to return the genes that produce differences in the samples with regard to the associated notation and pathways in which are implicated.

This case study looked at the differences in expressions that can occur in female patients younger than 50 years of age compared with those older than 50, since the latter group respond better to the treatments used.

This study is interesting because finding genes that behave differently can bring new information and the possibility of adjusting treatments for this type of cancer in younger patients.

The multi-agent system is developed in such a way that allows new agents to be inserted with new techniques or existing data to be modified for analysis. The system provides access to various databases so different cancer datasets can be introduced.

The system uses a CBR that handles all information obtained from the databases and allows the incorporation of new information that may be used in future analysis.

**Acknowledgments** This work has been c has been supported by the Spanish Government through the project iHAS (grant TIN2012-36586-C01/C02/C03) and FEDER funds.

## References

1. Argente, E., Botti, V., Carrascosa, C., Giret, A., Julian, V., Rebollo, M.: An abstract architecture for virtual organizations: The THOMAS approach. Knowl. Inf. Syst. **29**(2), 379–403 (2011)
2. Armstrong, N.J., Van De Wiel, M.A.: Microarray data analysis: From hypotheses to conclusions using gene expression data. Cell Oncol. **26**(5–6), 279–290 (2004)
3. Choon, Y.W., Mohamad, M.S., Deris, S., Illias, R.M., Chong, C.K., Chai, L.E.: A hybrid of bees algorithm and flux balance analysis with OptKnock as a platform for in silico optimization of microbial strains. Bioprocess Biosyst. Eng. **37**(3), 521–532 (2014)
4. Corchado, J.M., De Paz, J.F., Rodríguez, S., Bajo, J.: Model of experts for decision support in the diagnosis of leukemia patients. Artif. Intell. Med. **46**(3), 179–200 (2009)
5. De Paz, J.F., Bajo, J., Vera, V., Corchado, J.M.: MicroCBR: a case-based reasoning architecture for the classification of microarray data. Appl. Soft Comput. **11**(8), 4496–4507 (2011)
6. De Paz, J.F., Benito R., Bajo, J., Rodríguez-Vicente A., Abáigar M.: aCGH-MAS: analysis of aCGH by means of multi-agent system. Hindawi Publishing Corporation (In press)
7. Dermitzakis, E.T.: From gene expression to diseaserisk. Nat. Genet. **40**(5), 492–493 (2008)
8. Elzi, D.J., Song, M., Hakala, K., Weintraub, S.T., Shiio, Y.: Wnt antagonist SFRP1 functions as a secreted mediator of senescence. Mol. Cell. Biol. **32**(21), 4388–4399 (2012)
9. Finn, R.D., Tate, J., Mistry, J., Coggill, P.C., Sammut, S.J., Hotz, H.-R., Forslund, K., Eddy, S.R., Sonnhammer, E.L.L., Bateman, A.: The Pfam protein families database. Nucl. Acids Res. **36**(Database issue), D281–D288 (2008)
10. Gene Ontology Consortium.: The gene ontology (GO) database and informatics resource. Nucl. Acids Res. **32**(Database issue), D258–D261 (2004)
11. Joshi-Tope, G., Gillespie, M., Vastrik, I., D'Eustachio, P., Schmidt, E., de Bono, B., Jassal, B., Gopinath, G.R., Wu, G.R., Matthews, L., Lewis, S., Birney, E., Stein, L.: Reactome: a knowledgebase of biological pathways. Nucl. Acids Res. **33**(Database issue), D428–D432 (2005). doi:10.1093/nar/gki072
12. Kapur, K., Xing, Y., Ouyang, Z., Wong, W.H.: Exon arrays provide accurate assessments of gene expression. Genome Biol. **8**(5), R82 (2007)

13. Knudsen, S.: Cancer Diagnostics with DNA Microarrays. Wiley-Liss (2006)
14. Lockhart, D.J., Winzeler, E.A.: Genomics, gene expression and DNA arrays. Nature **405**, 827–836 (2000)
15. Maglietta, R., D'Addabbo, A., Piepoli, A., Perri, F., Liuni, S., Pesole, G., Ancona, N.: Selection of relevant genes in cancer diagnosis based on their prediction accuracy. Artif. Intell. Med. **40**(1), 29–44 (2007)
16. Mantripragada, K.K., Buckley, P.G., Diaz de Stahl, T., Dumanski, J.P.: Genomic microarrays in the spotlight. Trends Genet. **20**(2), 87–94 (2004)
17. Nguyen, T.P., Ho, T.B.: Detecting disease genes based on semi-supervised learning and protein–protein interaction networks. Artif. Intell. Med. **54**(1), 63–71 (2012)
18. Ng, S.-K., Zhang, Z., Tan, S.-H., Lin, K.: InterDom: a database of putative interacting protein domains for validating predicted protein interactions and complexes. Nucl. Acids Res. **31**(1), 251–254 (2003)
19. Nuber, U.A.: DNA Microarrays. Taylor & Francis group, New York (2005)
20. Pinkel, D., Albertson, D.G.: Array comparative genomic hybridization and its applications in cancer. Nat. Genet. **37**, 11–17 (2005)
21. The UniProt Consortium.: The universal protein resource (UniProt). Nucl. Acids Res. **35** (Database issue), D193–D197 (2007)
22. Vogelstein, B., Kinzler, K.W.: Cancer genes and the pathways they control. Nat. Med. **10**, 789–799 (2004)
23. Wang, P., Young, K., Pollack, J., Narasimham, B., Tibshirani, R.: A method for callong gains and losses in array CGH data. Biostatistics **6**(1), 45–58 (2005)
24. Ylstra, B., Van den Ijssel, P., Carvalho, B., Meijer, G.: BAC to the future! or oligonucleotides: a perspective for microarray comparative genomic hybridization (array CGH). Nucl. Acids Res. **34**, 445–450 (2006)
25. Zhang, J., Wu, L.-Y., Zhang, X.-S., Zhang, S.: Discovery of co-occurring driver pathways in cancer. BMC Bioinform. **15**(1), 271 (2014)

# Erratum to: Diagnostic Knowledge Extraction from MedlinePlus: An Application for Infectious Diseases

**Alejandro Rodríguez-González, Marcos Martínez-Romero, Roberto Costumero, Mark D. Wilkinson and Ernestina Menasalvas-Ruiz**

## Erratum to: Chapter 'Diagnostic Knowledge Extraction from MedlinePlus: An Application for Infectious Diseases' in: R. Overbeek et al. (eds.), *9th International Conference on Practical Applications of Computational Biology and Bioinformatics*, Advances in Intelligent Systems and Computing 375, DOI 10.1007/978-3-319-19776-0_9

In Chapter 'Diagnostic Knowledge Extraction from MedlinePlus: An Application for Infectious Diseases', in the chapter author affiliation Mexico should be replaced with "Spain".

The online version of the original chapter can be found under
DOI 10.1007/978-3-319-19776-0_9

A. Rodríguez-González (✉) · M.D. Wilkinson
Universidad Politécnica de Madrid – Centro de Biotecnología y Genómica de Plantas, Madrid, Spain
e-mail: alejandro.rodriguezg@upm.es

M.D. Wilkinson
e-mail: mark.wilkinson@upm.es

M. Martínez-Romero
Universidad de A Coruña – Centro IMEDIR, A Coruña, Spain
e-mail: marcosmartinez@udc.es

R. Costumero · E. Menasalvas-Ruiz
Universidad Politécnica de Madrid – Centro de Biotecnología Biomédica, Madrid, Spain
e-mail: roberto.costumero@upm.es

E. Menasalvas-Ruiz
e-mail: ernestina.menasalvas@upm.es

 
R. Overbeek et al. (eds.), *9th International Conference on Practical Applications of Computational Biology and Bioinformatics*, Advances in Intelligent Systems and Computing 375, DOI 10.1007/978-3-319-19776-0_16

# Author Index





R. Overbeek et al. (eds.), *9th International Conference on Practical Applications of Computational Biology and Bioinformatics*, Advances in Intelligent Systems and Computing 375, DOI 10.1007/978-3-319-19776-0

Printed in the United States
By Bookmasters